BIOCHEMISTRY RESEARCH TRENDS

INULIN

BIOCHEMISTRY, FOOD SOURCES AND HEALTH IMPLICATIONS

BIOCHEMISTRY RESEARCH TRENDS

Additional books in this series can be found on Nova's website
under the Series tab.

Additional e-books in this series can be found on Nova's website
under the e-book tab.

BIOCHEMISTRY RESEARCH TRENDS

INULIN

BIOCHEMISTRY, FOOD SOURCES AND HEALTH IMPLICATIONS

MICHAEL G. MARLOW
EDITOR

nova publishers
New York

Library of Congress Cataloging-in-Publication Data

ISBN: 978-1-63463-985-9

Library of Congress Control Number: 2014960321

Published by Nova Science Publishers, Inc. † New York

CONTENTS

Preface		**vii**
Chapter 1	Distinctive Viscoelastic Behaviour and Sensory Features Introduced by Inulin/Extra Virgin Olive Oil Blends in Mashed Potatoes: Effects of Ingredients Ratio and Freeze/Thaw Cycle *María José Jiménez, Wenceslao Canet and María Dolores Alvarez*	**1**
Chapter 2	Processing Plants Containing Inulin *Václav Veselý, Jana Budovičová, Jiří Hanika, Miroslav Punčochář and Martin Bárnet*	**57**
Chapter 3	Inulin: Modulation of Intestinal Microbiota and Propects in New Products *Marilyn Gonçalves Ferreira Kuntz, Ricardo Fernandes, Vinicius André do Rosário, Giovanna Medeiros Rataichesck Fiates and Erasmo Benicio Santos de Moraes Trindade*	**103**
Index		**147**

PREFACE

Inulin (INL) is produced by several different plants. Inulins are polysaccharides and are of interest for the development of healthy products because they simultaneously respond to a variety of consumer demands. It is fibre-enriched, prebiotic, low in fat and low in sugar. This book discusses the food sources, the biochemistry and health implications of inulin.

Chapter 1 - Inulin (INL) is a functional food ingredient and extra virgin olive oil (EVOO) has interesting nutritional characteristics. This chapter investigates the effect of the addition of INL and EVOO blends at different INL:EVOO ratios (0:0, 0:60, 15:45, 30:30, 45:15, 60:0, 30:45 and 45:30) on oscillatory shear experiments, instrumental colour and sensory properties (appearance, texture and taste attributes) of fresh and frozen/thawed mashed potatoes (FMP and F/TMP respectively). To ameliorate detrimental freezing effects, mashed potatoes (MP) were formulated without and with added hydrocolloids (MPA and MPB respectively). Dynamic shear data revealed weak gel-like behaviour in the MP samples, with magnitudes of storage modulus (G') higher than those of loss modulus (G''), and a frequency-related increase mainly in elasticity. INL and EVOO addition (either alone or blended) produced softer systems as compared to 0:0 controls, although the addition of EVOO alone (0:60 ratio) produced superior softening than that of only INL (60:0 ratio). FMP and MPA samples presented more rigid structures than their F/TMP and MPB counterparts, which is associated with amylopectin retrogradation in the case of the samples subjected to a freeze/thaw cycle. Superior water-holding capacity was also observed in the case of samples with added hydrocolloids. Linear discriminant analysis (LDA) showed that the effect of the INL:EVOO ratio on sensory attributes was more significant in the F/TMP samples than in their FMP counterparts, likely due to a higher degree of oil droplet coalescence caused by freezing. In both FMP and F/TMP products, LDA indicated that *ingredient taste* was the attribute that best discriminated between the INL/EVOO blends used, a finding which proves that in the case of EVOO, a liquid product, textural attributes are not as significant as those of aroma and taste. Categorical principal component analysis (CATPCA) showed that in the FMP samples, the first dimension was related with rough-smooth texture and the second with mechanical texture-appearance. In the F/TMP samples, the first dimension was associated with appearance (*authentic colour* and *uniformity*) and *ingredient taste* attributes, whereas the second was dominated by mechanical texture (*consistency* and *adhesiveness*). *Ingredient taste*, *consistency* and *adhesiveness* were the sensory attributes that showed a superior internal consistency (higher Cronbach alpha coefficient) in both FMP and F/TMP samples.

Chapter 2 - Sterile juice containing dissolved sacharides and polysacharides and filtration cake were obtained at sterilisation of raw juice and following crystallisation. This cake contained a precipitate with high content of vegetable proteines. Hydrolysis by hydrochloric acid was executed at two temperatures. Content of individual amino acids in hydrolysate was determined. In hydrolysate tryptofan was not present. Acidic hydrolysis of inulin was studied and a model describing dependence of the first order rate constant on temperature and pH of solution was proposed. The effect of heterogeneous ion exchange on hydrolysis of inulin was determined. On basis of the obtained results a method of electrolytic separation followed by hydrolysis of natural juice from tubers of yakon and JA was developed.

Fermentation experiments of hydrolysed juice of JA to ethyl alcohol were executed.

Chapter 3 - A healthy microbiota is defined as one that preserves and keeps it the well-being and the disease-free, especially of the gastrointestinal tract with three main functions: metabolism function, protection function and trophic role on the proliferation and differentiation of the intestinal epithelium and the development of the immune system modulation. However, some factors may influence this process of colonization, such as diet.

In recent decades, there has been a large increase in studies analyzing the differences of microbiota in healthy subjects and in various disease states. Inflammatory bowel disease, liver disease, diabetes and obesity were the most explored so far, with a much larger spectrum still being researched. To promote bifidogenic effect, the usual recommendation for supplementation of inulin-type prebiotics is a daily dose of 2.5-10 g.

Based on all the studies that show the importance of prebiotic inulin as an adjunct in the treatment of several related comorbidities, the industry has been using inulin to improve the sensory quality and nutritional value of food products, among them, there were baked food, meat, sweets and dairy products. And scientists have been working jointly so that new products are developed using inulin as prebiotic. However, performed work to evaluate, through clinical trials, the effect in vivo prebiotic product developed, few found jobs, mostly with dairy products thus identifying the need for more development work to investigate, in practical tests, the effect of inulin in many products currently developed by the industry.

In: Inulin
Editor: Michael G. Marlow
ISBN: 978-1-63463-985-9

Chapter 1

DISTINCTIVE VISCOELASTIC BEHAVIOUR AND SENSORY FEATURES INTRODUCED BY INULIN/EXTRA VIRGIN OLIVE OIL BLENDS IN MASHED POTATOES: EFFECTS OF INGREDIENTS RATIO AND FREEZE/THAW CYCLE

María José Jiménez[1], Wenceslao Canet[2] and María Dolores Alvarez[2,*]
[1]USAS, Sensory Analysis Service Unit
[2]Department of Characterization, Quality and Safety,
Institute of Food Science, Technology and Nutrition (ICTAN-CSIC),
Madrid, Spain

ABSTRACT

Inulin (INL) is a functional food ingredient and extra virgin olive oil (EVOO) has interesting nutritional characteristics. This chapter investigates the effect of the addition of INL and EVOO blends at different INL:EVOO ratios (0:0, 0:60, 15:45, 30:30, 45:15, 60:0, 30:45 and 45:30) on oscillatory shear experiments, instrumental colour and

* E-mail: mayoyes@ictan.csic.es.

sensory properties (appearance, texture and taste attributes) of fresh and frozen/thawed mashed potatoes (FMP and F/TMP respectively). To ameliorate detrimental freezing effects, mashed potatoes (MP) were formulated without and with added hydrocolloids (MPA and MPB respectively). Dynamic shear data revealed weak gel-like behaviour in the MP samples, with magnitudes of storage modulus (G') higher than those of loss modulus (G''), and a frequency-related increase mainly in elasticity. INL and EVOO addition (either alone or blended) produced softer systems as compared to 0:0 controls, although the addition of EVOO alone (0:60 ratio) produced superior softening than that of only INL (60:0 ratio). FMP and MPA samples presented more rigid structures than their F/TMP and MPB counterparts, which is associated with amylopectin retrogradation in the case of the samples subjected to a freeze/thaw cycle. Superior water-holding capacity was also observed in the case of samples with added hydrocolloids. Linear discriminant analysis (LDA) showed that the effect of the INL:EVOO ratio on sensory attributes was more significant in the F/TMP samples than in their FMP counterparts, likely due to a higher degree of oil droplet coalescence caused by freezing. In both FMP and F/TMP products, LDA indicated that *ingredient taste* was the attribute that best discriminated between the INL/EVOO blends used, a finding which proves that in the case of EVOO, a liquid product, textural attributes are not as significant as those of aroma and taste. Categorical principal component analysis (CATPCA) showed that in the FMP samples, the first dimension was related with rough-smooth texture and the second with mechanical texture-appearance. In the F/TMP samples, the first dimension was associated with appearance (*authentic colour* and *uniformity*) and *ingredient taste* attributes, whereas the second was dominated by mechanical texture (*consistency* and *adhesiveness*). *Ingredient taste*, *consistency* and *adhesiveness* were the sensory attributes that showed a superior internal consistency (higher Cronbach alpha coefficient) in both FMP and F/TMP samples.

Keywords: prebiotic; extra virgin olive oil; puréed cooked potatoes; descriptive analysis; viscoelasticity; freeze/thaw

INTRODUCTION

Considering the high competitiveness of today's markets, food products that are launched to the market place should be aligned with consumers' preferences as much as possible [1]. Functional product development provides an opportunity to contribute to the improvement of food quality and consumer health and well-being [2]. A common approach employed in the food industry to develop these kinds of added-value foods involves the addition of a healthy component. It is important to know to what extent this new component modifies the product, affecting its sensory properties. In the case of macromolecules such as polysaccharides or proteins, changes in composition can affect product structure in particular, modifying both rheological behaviour and perceived texture [3].

Inulin (INL) is of interest for the development of healthy products because it simultaneously responds to a variety of consumer demands: it is fibre-enriched, prebiotic, low fat and low sugar [4]. As a dietary fibre, INL passes through the digestive tract largely undigested. In the colon, it acts as a prebiotic because it is selectively fermented by the beneficial flora, stimulates their growth, and reinforces its action against putrefactive microorganisms [5]. The addition of INL to different foods has aimed to supplement them to increase fibre ingestion, in amounts that oscillate between 3 and 6 g per portion, or to assure its bifidogenic nature, adding 3-8 g per portion [6].

Olive oil has gained the interest of consumers because of its multiple culinary applications and nutritional content (high content of oleic acid and balance between monounsatured and polyunsaturated fatty acids) and its content of minor compounds (polyphenols and tocopherols that have antioxidant properties and volatile compounds that bestow an oil its unique flavour) [7]. On November 2004, the US Federal Drug Administration (FDA) permitted a claim on olive oil labels concerning: “the benefits on the risk of coronary heart disease of eating about two tablespoons (23 g) of olive oil daily, because of the monounsaturated fat (MUFA) in olive oil” [8]. These properties might fulfill consumer expectations regarding healthier, more nutritious, more flavourful, and more natural food even in countries outside the Mediterranean region where most olive oil is consumed [9]. The Int. Olive Council (IOC) defines extra virgin olive oil (EVOO) as the commercial grade for olive oil that has been extracted by physical processes alone, without the use of solvents or refining agents. The absence of refining processes helps EVOO to preserve its delicate aroma and nutritional properties that distinguish it from other edible oils [10].

Texture is one of the most important factors determining the quality/liking of food. Others are appearance, flavour, comprising taste and smell, and nutritional aspects. However, from a nutritional point of view it is the product composition and its digestibility that primarily matters. For these reasons food scientists and nutritionists like to be able to control the relation between product composition and sensory/texture perception [11].

In frozen vegetables, health quality, nutritional quality and aspects of sensory quality like colour and texture can be objectively assessed and controlled [12]. As the stimulus in texture perception is predominantly mechanical in nature, many instrumental methods are geared to determining rheological characteristics of the food product [13]. However, this does not guarantee matching the sensory perceived texture that only can be sensed and measured by the human being. Sensory research involves the study of texture perception using sensory panels and required input from e.g., psychology and neurophysiology [14].

A number of studies deal with the perception of semisolid mashed potatoes (MP) sensory attributes by using conventional techniques such as descriptive quantitative analysis (QDA) adapted from Adams et al. [15], and considering the evaluation of appearance, texture and taste attributes. The addition of xanthan gum (XG) clearly improved the taste of the MP subjected to a freeze/thaw cycle (F/TMP), whereas addition of amidated low methoxyl pectin at higher concentrations ($\geq$ 4.5 g kg^{-1}) produced an unpleasant taste [16]. Analogously, panellists scored both FMP and F/TMP samples with added dairy proteins unacceptable as regards pleasantness to taste [17]. In turn, in F/TMP samples, increased fibrousness was related to the addition of sodium caseinate, whereas improved creaminess was associated with XG addition [18]. In EVOO-enriched MP, panellists found that increasing yellowness improved the colour of the product, since the flesh of Kennebec potatoes is white and MP without added other ingredients look rather pale [19].

MPs behave like a high-volume-fraction dispersion, so that the addition of INL (extra dry matter) would be expected to result in a firmer product. There is little information on the sensory characteristics of INL and EVOO and how they are affected by possible interactions with other ingredients, like different kinds of hydrocolloids or starch. A complicating factor is that ingredients mostly affect several texture/sensory attributes at the same time [11]. EVOO is a liquid product, and therefore, upon addition of EVOO, the systems would be expected to be softer whatever the concentration used. This chapter was focused on the possibility of optimising MP formulation complying with an improved nutritional value and at the same time to gain more insights into the rheological and sensory properties of such systems.

The aim of this chapter was therefore: (1) to examine the effect of adding INL/EVOO blends at different INL:EVOO ratios on viscoelastic and sensory properties of FMP and F/TMP samples formulated without and with added hydrocolloids (designated MPA and MPB, respectively); (2) to use LDA to investigate the possibility of objective discrimination among the different performance conditions of MP samples on the basis of descriptive sensory properties; and (3) to reduce the number of sensory properties by means of linear combinations of these, using PCA. The reliability of the sensory attributes is also discussed in some detail.

MATERIALS AND METHODS

Materials and Preparation of MP Samples

The potatoes used were tubers (cv Kennebec) from Aguilar de Campoo (Burgos, Spain). INL (Orafti®HP, BENEO-Orafti, Tienen, Belgium) was a 'long-chain' INL with a degree of polymerisation, DP > 23 and purity of 99.5% (producer's data). EVOO (Carbonell, Sevilla, Spain) was chosen for addition to MP. Kappa-carrageenan (κ-C) (GENULACTA carrageenan type LP-60) and xanthan gum (XG) (Keltrol F [E]) were donated by Premium Ingredients, S.L. (Girona, Spain). I/EVOO blends were added to the samples at five different INL:EVOO ratios (15 g kg^{-1} (INL)/45 g kg^{-1} (EVOO), 15:45; 30 g kg^{-1} (INL)/30 g kg^{-1} (EVOO), 30:30; 45 g kg^{-1} (INL)/15 g kg^{-1} (EVOO), 45:15; 30 g kg^{-1} (INL)/45 g kg^{-1} (EVOO), 30:45; 45 g kg^{-1} (INL)/30 g kg^{-1} (EVOO), 45:30). Samples without added ingredients (0:0 control), with 60 g kg^{-1} added EVOO alone (0:60) and with 60 g kg^{-1} added INL alone (60:0) were also prepared for each type of MP and processing conditions. In this way, the effect of adding ingredients (0:0 vs. 0:60/15:45/30:30/45:15/60:0 or vs. 30:45/45:30), the effect of the variation of INL:EVOO ratio at a fixed total concentration (0:60/15:45/30:30/45:15/60:0) and the effect of either adding EVOO at a fixed INL concentration (30:30/30:45 or 45:15/45:30) or adding INL at a fixed EVOO content (15:45/30:45 or 30:30/45:30) can be analysed.

Tubers were manually washed, peeled and diced. MP samples were prepared in ~2000-g batches from 1185 g of potatoes, 450 mL of semi-skimmed in-bottle sterilised milk (fat content, 15.5 g kg^{-1}), 300 mL of water, 15 g of salt (NaCl) and the corresponding EVOO concentration (0-60 g kg^{-1}) using a TM 31 food processor (Vorwerk España, M.S.L., S.C., Madrid, Spain). INL (0-60 g kg^{-1}) was previously dissolved in 750 mL of water and milk at 70 °C for 15 min, agitating with a magnetic stirrer at 600 r.p.m. MP were prepared without and with added κ-C and XG (MPA and MPB samples, respectively). In the latter case, hydrocolloids (each at 1.5 g kg^{-1}) were added

Table 1. Composition and processing of the samples studied with codes used for their identification

INL ($g\ kg^{-1}$)	EVOO ($g\ kg^{-1}$)	κ-C + XG ($g\ kg^{-1}$)	Processing	Sample code
0	0	-	Fresh	FMPA-0:0
0	60	-	Fresh	FMPA-0:60
15	45	-	Fresh	FMPA-15:45
30	30	-	Fresh	FMPA-30:30
45	15	-	Fresh	FMPA-45:15
60	0	-	Fresh	FMPA-60:0
30	45	-	Fresh	FMPA-30:45
45	30	-	Fresh	FMPA-45:30
0	0	1.5 + 1.5	Fresh	FMPB-0:0
0	60	1.5 + 1.5	Fresh	FMPB-0:60
15	45	1.5 + 1.5	Fresh	FMPB-15:45
30	30	1.5 + 1.5	Fresh	FMPB-30:30
45	15	1.5 + 1.5	Fresh	FMPB-45:15
60	0	1.5 + 1.5	Fresh	FMPB-60:0
30	45	1.5 + 1.5	Fresh	FMPB-30:45
45	30	1.5 + 1.5	Fresh	FMPB-45:30
0	0	-	Frozen/thawed	F/TMPA-0:0
0	60	-	Frozen/thawed	F/TMPA-0:60
15	45	-	Frozen/thawed	F/TMPA-15:45
30	30	-	Frozen/thawed	F/TMPA-30:30
45	15	-	Frozen/thawed	F/TMPA-45:15
60	0	-	Frozen/thawed	F/TMPA-60:0
30	45	-	Frozen/thawed	F/TMPA-30:45
45	30	-	Frozen/thawed	F/TMPA-45:30
0	0	1.5 + 1.5	Frozen/thawed	F/TMPB-0:0
0	60	1.5 + 1.5	Frozen/thawed	F/TMPB-15:45
15	45	1.5 + 1.5	Frozen/thawed	F/TMPB-30:30
30	30	1.5 + 1.5	Frozen/thawed	F/TMPB-45:15
45	15	1.5 + 1.5	Frozen/thawed	F/TMPB-60:0
60	0	1.5 + 1.5	Frozen/thawed	F/TMPB-0:60
30	45	1.5 + 1.5	Frozen/thawed	F/TMPB-30:45
45	30	1.5 + 1.5	Frozen/thawed	F/TMPB-45:30

to the rest of ingredients in the form of a dry powder. The ingredients were first heated to 90 °C (17 °C min^{-1}) and kept at 90 °C for 30 min (blade speed:

40 r.p.m.) [20, 21]. Shearing was performed with a propeller. The amount of evaporated liquid was determined by weighing the ingredients before and after the first cooking and then replaced by adding milk. Next, all the ingredients were cooked for an additional 5 min at 90 °C. The mash was ground for 40 s (blade speed: 1200 r.p.m.) and 20 s (blade speed: 2600 r.p.m.), and then homogenised immediately through a stainless steel sieve (diameter: 1.5 mm). Two batches were continuously being prepared and blended and half of each fresh blend (FMP samples) was analysed immediately whilst the other half was subjected to a freeze/thaw cycle (F/TMP samples). Each MP composition was prepared twice but in different weeks to assure the appropriate experiment randomization.

Freezing, Thawing and Heating Processes

Following their preparation, MP samples were placed on flat freezing and microwave thawing trays and then frozen by forced convection with liquid nitrogen vapour in an Instron programmable chamber (model 3119-05, −70/+250 °C) at −60 °C until their thermal centres reached −24 °C [22]. After freezing, the samples were packed in polyethylene plastic bags, sealed under light vacuum (−0.05 MPa) on a Multivac packing machine (Sepp Haggenmüller KG, Wolfertschwenden, Germany), and placed in a domestic freezer for storage at −24 °C. Packed frozen samples with the plastic bags perforated were then thawed in a Samsung M1712N microwave oven (Samsung Electronics S.A., Madrid, Spain) by heating for 20 min at an output power rating of 600 W. After thawing, the temperature reached at the product thermal centre was measured in all cases (+85 ± 3 °C). Samples were brought to 55 °C by placing them in a Hetofrig CB60VS water-bath (Heto Lab Equipment A/S, Birker∅d, Denmark). The sample testing temperature was maintained at 55 °C as this is the preferred temperature for consumption of MP [19-23].

Rheological Characterisation. Oscillatory Dynamic Tests

A Bohlin CVR 50 controlled stress rheometer (Bohlin Instruments Ltd., Cirencester, UK) was used to conduct small amplitude oscillatory shear (SAOS) experiments using a plate-plate sensor system with a 2 mm gap (PP40, 40 mm) and a solvent trap to minimise moisture loss during tests. After loading the sample, there was a 5-min waiting period to allow the sample to recover and reach 55 °C. Temperature control at 55 °C was achieved with a Peltier Plate system (-40 to +180 °C; Bohlin Instruments). In order to determine the linear viscoelastic (LVE) region, the first stress sweeps were run at a constant frequency (ω) of 1 rad s^{-1} over a shear stress range of 3-300 Pa. The LVE range was limited to that amplitude range for which the complex modulus (G^*) was constant [24]. Loss tangent ($\tan\delta$ = G''/G'), storage modulus (G', Pa) and loss modulus (G'', Pa) values were also registered within the LVE region. The G' represents the non-dissipative component of mechanical properties and is characteristic of elasticity, while G'' represents the dissipative component of the mechanical properties and is characteristic of viscous flow [25]. Next three frequency sweeps were performed over the ω range of 0.1-100 rad s^{-1}, and again the $\tan\delta$, G' and G'' values were registered at 1 rad s^{-1}. Given that the appearance of the data on logarithmic data was nearly linear, a power law model was used to characterise the frequency dependence of both moduli [24, 26], from the following equations:

$$G' = G_0' \cdot \omega^{n'}$$

$$G'' = G_0'' \cdot \omega^{n''}$$

Where G_0' and G_0'' are elastic and viscous moduli at 1 rad s^{-1} respectively, and exponents n' and n'' denote the influence degree of ω on both moduli.

As a new sample was used each time for the dynamic tests, the resulting values were average values of the four determinations.

Colour Measurements

The colour of the MP samples in the pots was measured using a HunterLab model D25 (Reston, VA, USA) colour difference meter fitted with a 5-cm-diameter aperture. Results were expressed in accordance with the CIELAB system with reference to illuminant D65 and a visual angle of 10°. The parameters determined were L^* (L^* = 0 [black] and L^* = 100 [white]), a^* ($-a^*$ = greenness and $+a^*$ = redness), b^* ($-b^*$ = blueness and $+b^*$ = yellowness), C^* chroma (saturation) and H^* hue, as defined by the following equations:

$$C^* = (a^{*2} + b^{*2})^{1/2} \quad H^* = \arctan (b^* / a^*)$$

The colour was expressed as L^*/b^*, i.e., the white/ yellow ratio [17]. The total colour difference (ΔE^*) between the fresh control made without either added hydrocolloids or ingredients and the rest of the MP samples was calculated as follows:

$$\Delta E^* = \left[\left(\frac{L_n^* - L_1^*}{K_L\ S_L} \right)^2 + \left(\frac{C_n^* - C_1^*}{K_C\ S_C} \right)^2 + \left(\frac{H_n^* - H_1^*}{K_H\ S_H} \right)^2 \right]^{1/2}$$

Where L_1^*, C_1^* and H_1^* are the reference (fresh sample) colour parameter values and $S_L = 1.0$, $S_C = 1 + 0.045C_1^*$ and $S_H = 1 + 0.015C_1^*$. K_L, K_C, and K_H are parametric factors and vary with the experimental conditions. All were given a fixed unit value for the combination of our specific reference conditions [20].

Sensory Analysis

Samples were subjected to a quantitative descriptive analysis (QDA) adapted for each product [15]. Sensory evaluation was conducted by an eight-member panel with 9 years'specific training in potato purées [16-23]. Each sample was tested twice and average scores calculated, so that each sample was tested sixteen times in all. Panellists were asked to evaluate samples for appearance [authentic colour (Aucol), off-colours (Offcol), shine (Shi) and uniformity (Unicol)], texture [consistency (Con), adhesiveness (Adh), creaminess (Cre) and fibrousness (Fib)], and taste [sweetness (Swe), authentic taste (Autas), off-taste (Offtas) and ingredient taste (Ingtas)]. Scores were awarded on a scale of 1–5, in which 1 indicated total absence of the sensory attribute and 5 a very definite attribute.

Samples were evaluated, in duplicate, in morning sessions (11:00 a.m.-1:00 p.m.). Daily for 64 days, assessors were given one sample (about 50 g), for scoring the sensory attributes of the 32 total samples (FMPA, FMPB, F/TMPA and F/TMPB products with added 0:0, 0:60, 15:45, 30:30, 45:15, 60:0, 30:45 and 45:30 ratios). The corresponding controls (FMPA-0:0, FMPB-0:0, F/TMPA-0:0 or F/TMPB-0:0 samples) were used for reference together with the evaluated sample every day. All the samples were served in white plastic vessels at a temperature of 55 ± 1 °C, which was reached and kept constant by placing the product in a Hetofrig CB60VS (Heto Lab Equipment A/S, Birkerød, Denmark) water bath prior to testing.

Statistical Analysis

A three-way ANOVA with interactions was applied to evaluate how (INL + EVOO) blend, presence or absence of hydrocolloids and performance or not of a freeze/thaw cycle affected the oscillatory rheological properties and characteristics derived, and the colour of the MP. Minimum significant differences were calculated using Fisher's least significant difference (LSD) tests with a 99% confidence interval for the comparison of instrumental parameters.

Sensory attributes from the quantitative descriptive analysis (QDA) (discrete values) were subjected to Friedman two-factor ranked analysis of variance (k = 16 and N = 8; FMP and F/TMP samples were analysed separately), and where differences existed between samples a regular Wilcoxon signed rank test was used to pinpoint them [27]. As recommended by the ISO 8587 [28], the Friedman test was considered to establish the statistical significance of the differences that the panellists might detect between samples. Linear discriminant analysis (LDA) was performed on the FMP and F/TMP samples considered separately (n = 128) to determine variability across the different sensory attribute scores in terms of addition of hydrocolloids and (INL + EVOO) blend.

In the case of highly correlated data, typical of sets of food-quality measurements, Principal component analysis (PCA) may be utilised to reduce the data set into a smaller number of principal components that explain *most* of the inter-relationships among the variables in the data. In the situation where some or all of the variables are categorical, rather than 'standard' PCA, the more appropriate technique is categorical principal component analysis (CATPCA). CATPCA analyses were conducted on the FMP and F/TMP samples considered separately (n = 128) to investigate the interdependence of the sensory attributes via identification of new, uncorrelated variables.

Analogously, the Cronbach's coefficient α was used as reliability test [29], which was performed on the FMP and F/TMP samples considered separately (n = 128) to establish the levels of reability of the scores for each sensory property evaluated. All the mentioned analyses were performed using SPSS 19.0 software (SPSS, Inc., Chicago, IL, USA).

RESULTS AND DISCUSSION

Dynamic Rheological Properties

The common tests performed using small-amplitude oscillatory shear analyses are stress sweeps (for controlled stress rheometers), frequency sweeps, time sweeps, and temperature sweeps [30]. To better understand the structural changes that took place in the INL-EVOO-based MP structures, the influence of frequency in their viscoelastic properties was studied at 55 °C. Prior to the measurement of the mechanical spectra, it was necessary to determine the LVE region (i.e., the range in which *G*' and *G*" are independent of the stress/strain of the oscillation) [31]. As a general rule for solutions and gels, *G*' departs from linearity sooner than *G*"; therefore, to determine the LVE region, the component elastic was measured at a constant frequency of 0.16 Hz. The length of the LVE region varied with the composition of the samples, especially with the presence or absence of hydrocolloids and the performance or not of a freeze/thaw cycle, but in all the samples, a strain amplitude lower than 0.5% was comprised within the LVE region. Thus, frequency sweeps were performed at <0.5%. As an example, Figure 1 shows the stress sweeps of three FMPA samples. Dynamic curves of storage modulus *G*' (elastic component) are presented as functions of strain γ over four decades of strain. Above the LVE limit, *G*' decreased rapidly in all samples, indicating that the structure was highly prone to deformation. This is an appropriate test for analysing the gel character (*G*' > *G*") of samples, since as long as the strain amplitudes are below the limiting value (γ_{max}) the G^* [measurement of the overall resistance (elastic and viscous)] pattern has a plateau value, indicating that the gel structure is stable under these conditions [32].

The frequency dependence of *G*' and *G*" (mechanical spectra) of the different INL-EVOO-based MP samples before and after the freeze/thaw cycle is shown in Figure 2.

A material that is frequency independent over a large time scale range is solid-like; a true gel is such a material. In contrast, strong frequency dependence suggest a material structure with molecular entanglements that behaves more like a solid at higher frequencies and more like a liquid at lower frequencies [33]. The difference between the strong (true) gel and the weak gel can be established using mechanical spectroscopy [32, 34]. In the former, the molecular rearrangements within the network are reduced over the time scales analysed; *G*' is higher than *G*" and almost independent of ω (low and similar *n*' and *n*" parameters).

In this chapter, the mechanical spectra revealed the existence of soft gels, with values of *G*' higher than values of *G*" and with a weak dependence with frequency throughout the frequency range measured (Figure 2). Such behaviour has been reported elsewhere in tomato and vegetable purée-based baby foods [35, 36]. In each food, the rheological behaviour is directly related to its formulation; the conformational changes experienced by potato starch were largely responsible for the predominantly elastic behaviour of the systems. However, significant differences were found between the viscoelastic properties and rheological characteristics recorded for MP made under different conditions. It is then observed that the rheology of these mixtures is governed by a continuous phase (amylose/amylopectin matrix), and eventually modify by the presence of INL/EVOO blends. In FMPA, FMPB and F/TMPB products (Figures 2a, 2b, 2d, respectively), the sample with no (INL+EVOO) blends exhibited greater elasticity than those with added INL/EVOO content. This indicates a shift in the viscoelastic behaviour of the samples with added INL:EVOO ratios towards more viscous-like products. In F/TMPA systems (Figure 2c), the sample with 60 g kg^{-1} added INL had significantly higher *G*' value than the 0:0 control. González-Tomás et al. [31] reported that by adding 7.5% of long-chain INL to low-fat starch-based custard desserts they obtained viscoelastic properties similar to the whole-milk sample.

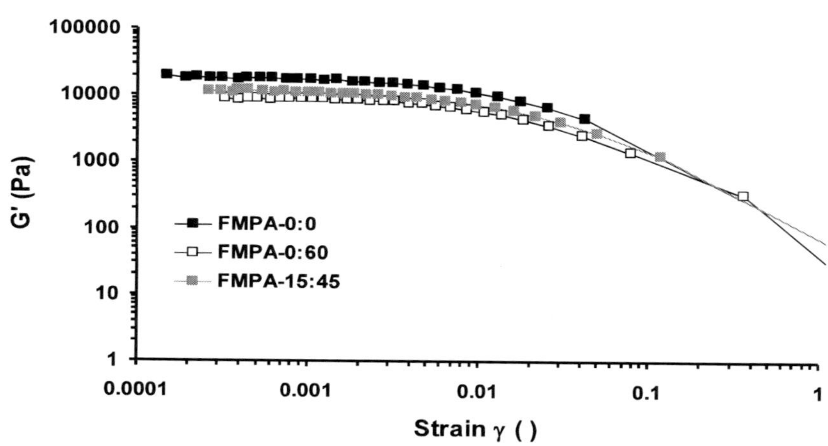

Figure 1. Stress sweeps of three freshly prepared mashed potatoes without added hydrocolloids and with added INL:EVOO ratios of 0:0, 0:60 and 15:45.

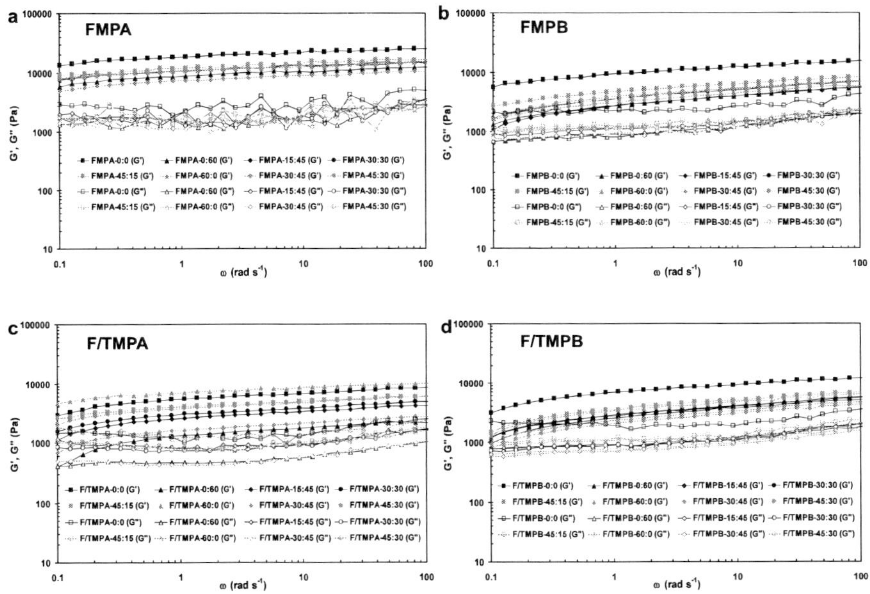

Figure 2. Dynamic properties, storage modulus (*G*', Pa) and loss modulus (*G*", Pa) versus frequency (rad s^{-1}) for fresh (FMP) and frozen/thawed mashed (F/TMP) potatoes with added different INL:EVOO ratios (0:0, 0:60: 15:45, 30:30, 45:15, 60:0, 30:45 and 45:30). (a) FMP without added hydrocolloids (FMPA). (b) FMP with added hydrocolloids (FMPB). (c) F/TMP without added hydrocolloids (F/TMPA). (d) F/TMP with added hydrocolloids (F/TMPB).

Frozen/thawed counterparts exhibited lower elasticity (Figures 2c, 2d). In contrast, Navarro et al. [24] found higher *G*' values in slowly frozen corn starch pastes compared with unfrozen and with quickly frozen samples. The slowly frozen starch systems showed also lower resistance to the applied stress despite their higher rigidity. The effect of freezing rate was explained due to the fact that the slow decrease in temperature gave amylose molecules higher opportunity to align and to release water trapped within the amylose network, favouring amylose retrogradation. In addition, slower freezing rates favoured large ice crystal formation, which distorted the system structure and contributed to thickening of the amylose network. In this chapter, both XG and κ-C are added to the MP to ameliorate the detrimental effects (syneresis, amylose retrogradation and rheological changes) of a freeze/thaw cycle [21]. Analoguosly, the addition of small amounts of XG to white sauces made with starches from different botanical sources significantly improved freeze/thaw stability [37]. According to the latter authors, this was probably due to XG interaction with solubilised amylose, reducing amylose-amylose interactions and therefore the extent of retrogradation. However, it was also reported that XG does not prevent ice recrystallisation or amylopectin retrogradation [38]; the fact that XG fails to prevent the latter is attributed to the fact that the gum acts outside the starch granule while amylopectin retrogradation takes place in the granule. Therefore, the decrease in the *G*' and *G*" values detected in the F/TMP samples as compared to their FMP counterparts could be attributed to amylopectin retrogradation associated with the freeze/thaw cycle.

Moreover, both FMPA and F/TMPA samples were more elastic and presented a significantly higher magnitude of *G*' than the samples with added hydrocolloids. The storage modulus of the matrix is attenuated by the presence of κ-C and XG, the effect being stronger after a freeze/thaw cycle. Note as in FMPA samples (Figure 2a), some oscillation in *G*" data may be appreciated, which could indicate a significant effect of slippage on viscosity in freshly prepared MP samples formulated without added hydrocolloids. In contrast, hydrocolloids addition attenuated the differences in the loss modulus of the MP samples made with added INL:EVOO ratios (Figures 2b and d). Rodríguez-Hernández and Tecante [39] found that xanthan exerted a greater effect on *G*", while *i*-carrageenan exerted a greater effect on *G*'. This result could be attributed to the fact that XG does not form gels and therefore its presence has a greater impact on the viscose rather than on the elastic response.

In turn, Table 2 shows the results of three-way ANOVA for dynamic rheological properties (tanδ, *G*' and *G*") and rheological characteristics (*G*'/*G*" and slopes of storage and loss moduli (*n*' and *n*", respectively)) with INL:EVOO ratio, hydrocolloid addition and a freeze/thaw cycle as the main effects. For the confidence level of 99% ($P < 0.01$), significant differences were found between means due to INL:EVOO ratio and hydrocolloid addition for all the dynamic properties and rheological characteristics, whereas only nonsignificant differences were found between means due to a freeze/thaw cycle for *n*". The average value of the *G*'/*G*" ratio was close to 5 for the MPA and FMP samples and nearly 3 for the MPB and F/TMP ones, supporting that the fresh product without added hydrocolloids behaves like a stronger gel structure (Figure 2a). On the other hand, the samples with higher INL content (ratios 45:15 and 60:0) presented higher *G*'/*G*" ratio values, although these values were similar to that of the control without added ingredients. However, samples with higher EVOO concentrations (ratios 0:60 and 30:45) had lower *G*'/*G*" ratio values. This result is to be expected as increasing concentrations of liquid oil are added to the product, increasing the oil-phase volume fraction. In oil-in-water emulsions, the extent of the linear region decreased with increasing oil-phase volume fraction from 20 to 40% v/v [40]. For their part, Dickinson and Chen [41] suggested that oil/water emulsions may undergo a behaviour transition from predominantly entropic behaviour to predominantly enthalpic behaviour, with increasing oil-phase volume fraction.

In order to rationalise these results, it may be thought that the amylose/amylopectin constitutes a matrix, in which INL crystallites and oil droplets would act as filler particles. This speculation is in agreement with the microstructure observed in INL-EVOO-based MP samples [42], where amylose/amylopectin forms a rigid matrix in which INL particles are trapped. Moreover, in this framework, INL crystallites should be regarded as inactive fillers, as they contribute negatively to the viscosity of continuous phase (somehow inhibiting the cross-linking between amylose/amylopectin chains, and hence weakening the polysaccharide matrix). This behaviour is typical of gels filled with deformable particles [43]. According to the latter authors, in gels containing deformable particles, the linear decrease in *G*' in line with increasing volume fractions is due to particle compliance under stress or to particle separation from the matrix, thereby causing gel weakening. Zimeri and Kokini [44] also observed the presence of INL crystals, by optical microscopy, which were immersed in the continuous phase in samples with 5% long-chain INL. They postulated that these crystals may be the result of recrystallisation or decreased INL solubility due to cooling the hot dispersions. In turn, Bayarri et al. [45] observed by SEM, the presence of some INL aggregates embedded in the continuous matrix of starch-dairy systems, which were attributed to recrystallisation mechanism.

The influence of particle-matrix interactions was also studied by van Vliet [46] who found that a linear decrease in gel strength accompanied by an increase in volume fractions only occurred with non-interacting gel materials. This finding was attributed to the formation of aqueous boundary layers around each particle so that there was no adhesion between the particle and the matrix. As stress is applied to the system, small amounts of water may be released from the INL crystallites, thus forming an aqueous boundary layer around them. This layer would reduce any interactions with the amylose/amylopectin matrix and lead to the formation of weak points in the gelatinised starch gel. Consequently, when these gels suffer small deformations, they behave as if filled with particles with the rheological properties of water.

Table 2. Effects of INL:EVOO ratio, hydrocolloid addition and a freeze/thaw cycle on viscoelastic properties, rheological characteristics and colour parameters of MP samples

Source	tanδ	G' (Pa)	G'' (Pa)	G'/G''	n'	n''	L^*/b^*	ΔE^*
Main effects:								
A: INL:EVOO ratio								
0:0	0.24_e	10144.00_a	1974.00_a	4.66_a	0.127_e	0.077_f	9.94_b	1.28_g
0:60	0.32_a	3848.31_f	914.22_{de}	3.65_e	0.166_a	0.121_a	5.57_f	8.02_a
15:45	0.28_{bc}	4806.69_e	1019.92_{cd}	3.97_{bc}	0.159_b	0.113_{bc}	6.12_e	5.90_b
30:30	0.29_{bc}	5205.44_d	1126.56_c	3.95_{cd}	0.151_c	0.106_{cd}	6.80_d	4.38_c
45:15	0.26_{de}	5932.12_c	1266.54_b	4.45_a	0.135_d	0.085_e	8.03_c	2.80_f
60:0	0.25_e	6303.31_b	1276.99_b	4.65_a	0.137_d	0.087_e	13.09_a	3.06_e
30:45	0.30_b	3580.25_f	874.36_e	3.71_{de}	0.157_b	0.114_b	6.10_e	5.74_b
45:30	0.27_{cd}	5085.56_{de}	1086.20_c	4.19_b	0.148_c	0.010_d	6.87_d	3.96_d
P values	<0.001	<0.001	<0.001	<0.001	<0.001	<0.001	<0.001	<0.001
LSD (99%)	*0.019*	*364.929*	*122.532*	*0.239*	*0.009*	*0.008*	*0.088*	*0.200*
B: Hydrocolloid addition								
Without κ-C and XG	0.22_b	7379.97_a	1229.07_a	5.14_a	0.112_b	0.082_b	8.13_a	3.45_b
With κ-C and XG	0.33_a	3846.45_b	1155.63_b	3.16_b	0.183_a	0.118_a	7.50_b	5.33_a
P values	<0.001	<0.001	0.0014	<0.001	<0.001	<0.001	<0.001	<0.001
LSD (99%)	*0.010*	*182.465*	*61.266*	*0.120*	*0.004*	*0.004*	*0.044*	*0.100*
C: A Freeze/thaw cycle								
Fresh	0.24_b	7728.97_a	1470.31_a	4.84_a	0.132_b	0.100_a	7.90_a	4.57_a
Frozen/thawed	0.31_a	3497.45_b	914.39_b	3.47_b	0.163_a	0.100_a	7.73_b	4.22_b
P values	<0.001	<0.001	<0.001	<0.001	<0.001	0.972	<0.001	<0.001
LSD (99%)	*0.010*	*182.465*	*61.266*	*0.120*	*0.004*	*0.004*	*0.044*	*0.100*
Interactions								
AB	<0.001	<0.001	<0.001	<0.001	<0.001	<0.001	<0.001	<0.001
AC	<0.001	<0.001	<0.001	<0.001	<0.001	<0.001	<0.001	<0.001
BC	<0.001	<0.001	<0.001	<0.001	<0.001	<0.001	<0.001	<0.001

INL, inulin; EVOO, extra virgen olive oil; κ-C, kappa-carrageenan; XG, xanthan gum.
abcdefgDifferent lower case letters in the same column and factor studied indicate significant diferences ($P < 0.01$).

In turn, Table 2 shows that the magnitudes of the resulting straight lines were small, and in all cases *n*" values were lower than *n*' values, showing that *G*' was more frequency-dependent than *G*" (Figure 2). The higher *n*' and *n*" values were obtained for samples containing 60 g kg^{-1} EVOO, indicating that these systems possessed networks which were transient in time and involved specific interactions between denser and less flexible particles, such as droplet aggregates. The MP samples presented here contain either Na^+, from salt incorporation, or Ca^{2+} ions, proceeding mainly from added milk. One of the possible effects of Ca^{2+} ions is the formation of droplet aggregates through some Ca^{2+}-mediated bridging, as suggested by Sosa-Herrera et al. [47]. These aggregates could further disturb the formation of the polysaccharide matrix.

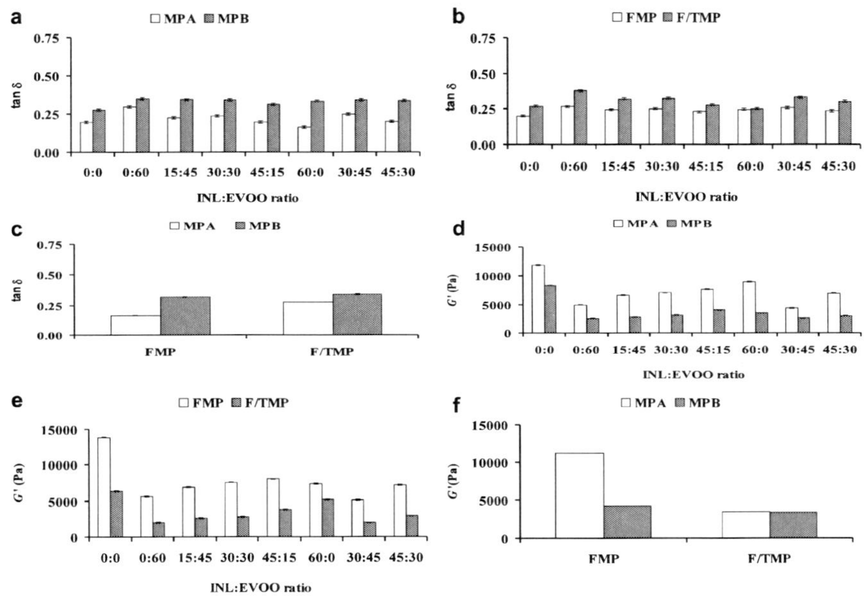

Figure 3. Rheological properties of MP samples: (a), (b), (c) loss tangent; (d), (e), (f) storage modulus. MPA: mashed potatoes without added hydrocolloids; MPB: mashed potatoes with added hydrocolloids; FMP: fresh mashed potatoes; F/TMP: mashed potatoes subjected to a freeze/thaw cycle; INL: inulin; EVOO: extra virgin olive oil.

In addition, the three binary interactions between factors had a significant effect on all the dynamic properties and rheological characteristics, meaning that the effect of the INL:EVOO ratio on the rheological properties was dependent on the presence or absence of hydrocolloids and the performance or not of a freeze/thaw cycle. Graphs of the dynamic functions with different binary interactions are shown in Figures 3 and 4. The graph of variation in the tanδ value based on the INL:EVOO ratio for both MPA and MPB samples (Figure 3a) shows that, in all cases, tanδ was lower in the MPA than in the MPB samples. Therefore, the latter samples were softer because of XG-water interactions, which caused a decrease in starch-alone viscosity, confirming previous findings [26]. All the MPB samples had significantly higher tanδ

values than the 0:0 control. As a result, addition of INL and/or EVOO (either alone or blended) produced softer, liquid-like systems. Note that the presence of hydrocolloids in the systems rendered the effect of the variation of the INL:EVOO ratio on the viscoelasticity in the MPB samples less appreciable. In the MPA products, the lowest tanδ value with respect to 0:0 control was recorded at a ratio of 60:0, indicating that in these systems the addition of INL alone gave rise to an improvement of solid character. This reinforcement of the structure could be attributed to the exclusion effect of the starch, which would lead to the formation of aggregates containing INL crystals [45].

The INL:EVOO ratio-dependent variation in tanδ value in both FMP and F/TMP samples is shown in Figure 3b. Except at a ratio of 60:0, FMP samples showed significantly lower tanδ values than their F/TMP counterparts; this could be ascribed to phase separation caused by retrogradation of the starch occurring in the frozen/thawed MPA systems [48]. In the F/TMP products, the variation of the INL:EVOO ratio at a fixed total concentration (60 g kg^{-1}) showed that when EVOO decreased and INL increased (0:60, 15:45 and 30:30 ratios) the products behaved like more solid systems, although there were nonsignificant differences in the loss tangent values of the samples with added 60:0 and 0:0 ratios (Figure 3b). Guggisberg et al. [49] observed that firmness and creaminess of low-fat set yogurt increased with both INL and fat content but the highest level of INL considered (4%) was not enough to imitate the whole-milk yogurt. In the FMP systems, all samples with added ingredients had significantly higher tanδ values than 0:0 control. In turn, the processing-dependent variation in tanδ value for MPA and MPB samples is shown in Figure 3c. A freeze/thaw cycle did not significantly affect sample viscoelasticity in the MPB samples but significantly increased it in MPA ones, corroborating the ability of κ-C and XG to provide freeze/thaw stability [21, 22, 26]. This fact was probably due to hydrocolloid interaction with solubilised amylose, reducing amylose-amylose interactions and therefore the extent of retrogradation [37].

The graph of variation in the *G*' value based on the INL:EVOO ratio for both MPA and MPB samples (Figure 3d) shows that, in all cases, elasticity was lower in the MPB than in the MPA samples. The probable explanation lies in the XG-water interactions that take place in the samples containing hydrocolloids [20, 21]. MPB samples occlude a great amount of water, which results in less shear thinning [42]. All the other samples (both MPA and MPB) had significantly lower values of *G*' with respect to 0:0 control. In both MPA and MPB products, the decrease in *G*' with respect to 0:0 ratio samples was greater when either 0:60 or 30:45 ratios were added, again indicating that addition of higher EVOO content caused more softening. At a fixed total concentration (60 g kg^{-1}), the elasticity increased linearly with increasing INL and decreasing EVOO concentrations. In the MPA products and at a fixed total concentration, the lowest and highest *G*' values were registered in samples with added 0:60 and 60:0, respectively. Also, at a fixed EVOO concentration, MPA sample elasticity decreased with INL content (15:45 vs. 30:45), whereas at a fixed INL concentration, elasticity significantly decreased with increasing EVOO concentration (30:30 vs. 30:45 and 45:15 vs. 45:30). These results suggest that EVOO was a more significant factor in the decrease in elasticity of the systems than INL.

FMP samples showed significantly higher *G*' values than their F/TMP counterparts (Figure 3e), which is ascribed to starch retrogradation occurring in the MPA samples subjected to a freeze/thaw cycle. In both FMP and F/TMP products, the variation of the INL:EVOO ratio at a fixed total concentration showed that when EVOO decreased and INL increased (0:60, 15:45, 30:30, 45:15 and 60:0 ratios or 30:45 and 45:30 ratios) the products behaved like more solid systems. This result is likely related to the formation of small insoluble INL particles. Such small particles can be produced by controlling heating and cooling of the dissolved INL to induce nucleation and thus small insoluble crystal formation [50]. Franck and De Leenheer [51] reported that at a high concentration (>15% for long-chain INL), INL forms a network composed of solid crystalline particles after shearing which confers gel-like texture to it. *G*' values seem to indicate that, in the MP systems, the presence of INL at quite high concentration ($\geq$ 45 g kg^{-1}) favoured the development and manifestation of its gelling properties. A freeze/thaw cycle significantly decrease *G*' in both MPA and MPB samples but the processing had a much less significant effect on the elasticity in the systems containing hydrocolloids (Figure 3f), again confirming the freeze-thaw stability conferred by the κ-C/XG blend [20, 42].

The variation in the *G*" values with the INL:EVOO ratio was very similar to that observed in the values of *G*' (Figure 4). The chief point of note is that differences in the viscosity of MPA samples with respect to their MPB counterparts (Figure 4a) were much less significant than those found for the elasticity (Figure 3d). On the other hand, at a ratio of 0:0, MPB samples showed significantly higher *G*" values than their MPA counterparts, again reflecting that in absence of INL/EVOO blends, xanthan exerted a greater effect on *G*" than on *G*' [21]. Analogously, a freeze/thaw cycle significantly decreased viscosity in both MPA and MPB samples, although the decrease was much less significant in the samples with added hydrocolloids (Figure 4c).

This behaviour can be explained taking account that XG influenced the *G*" component more and the networks that are formed when solutions of XG are frozen and thawed are much stronger and more cohesive [52].

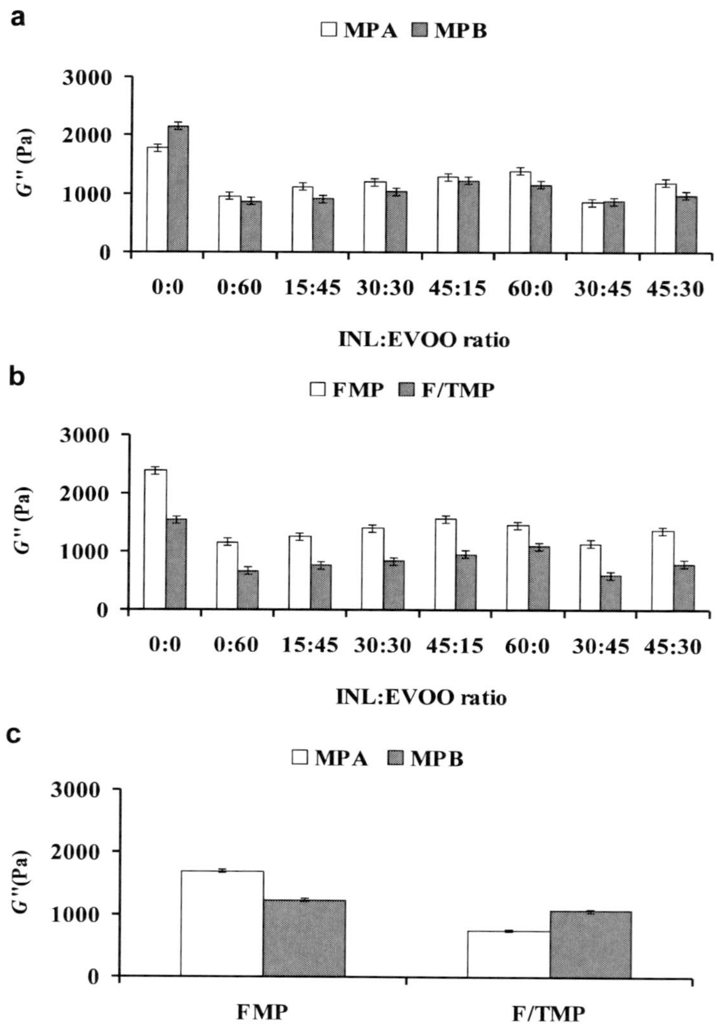

Figure 4. Rheological properties of MP samples: (a), (b), (c) loss modulus. MPA: mashed potatoes without added hydrocolloids; MPB: mashed potatoes with added hydrocolloids; FMP: fresh mashed potatoes; F/TMP: mashed potatoes subjected to a freeze/thaw cycle; INL: inulin; EVOO: extra virgin olive oil.

Instrumental Colour Measurements

Effects of INL:EVOO ratio, hydrocolloid addition and a freeze/thaw cycle on the values of the white/yellow ratio (L^*/b^*) and the total colour difference (ΔE^*) of the MP samples are also shown in Table 2. All the three main factors and their binary interactions had a significant effect on both L^*/b^* and ΔE^*. The variation in the L^*/b^* value based on INL:EVOO ratio for both MPA and MPB samples (Figure 5a) shows that in both cases, the lowest and highest L^*/b^* values were registered in the samples containing 0:60 and 60:0 ratios, respectively. Thus, addition of either EVOO alone or blended with INL produced a darker-coloured mashed potato (lower L^*/b^* ratio) than the 0:0 control. An increase in EVOO level favours either a higher L^* value (lightness) because of an increase in the overall light scattering associated with the scattering properties of fat [53], as well as a higher b^* value because of an increase in chloroplast pigment (chlorophyll and carotenoids) content. The pigment profile of the virgin olive oil comprises chlorophyll a, chlorophyll b and derivative pigments associated with the acidic medium of the oil extraction process [54]. The influence of droplet characteristics on the optical properties of coloured oil-in-water emulsions was examined [53]. The lightness (L^*) of the emulsions increased with increasing droplet concentration and decreasing droplet size. As the droplet concentration increases so does the reflectance because the droplets scatter light more effectively and hence the light beam is unable to penetrate further into the product and be absorbed.

In addition, in both MPA and MPB samples and at a fixed total concentration, L^*/b^* increased linearly with increasing and decreasing INL and EVOO contents, respectively (0:60, 15:45, 30:30, 45:15, 60:0). Unlike EVOO, INL produced a lighter-coloured mashed potato (higher L^*/b^* ratio). Commercially available INL is a spray-dried white powder which gives an opaque gel at higher concentrations when mixed properly with water [55]. In the case of INL/EVOO blends addition, except for the lowest EVOO content (60:0 ratio), all the other samples had significantly lower L^*/b^* values than the 0:0 control (Figure 5a), indicating that the darkening produced by EVOO addition is stronger than the lightening caused by INL incorporation. With regard to 0:0 control, increase yellowness caused by EVOO addition improved the product colour, as the flesh of Kennebec is too white [23, 42].

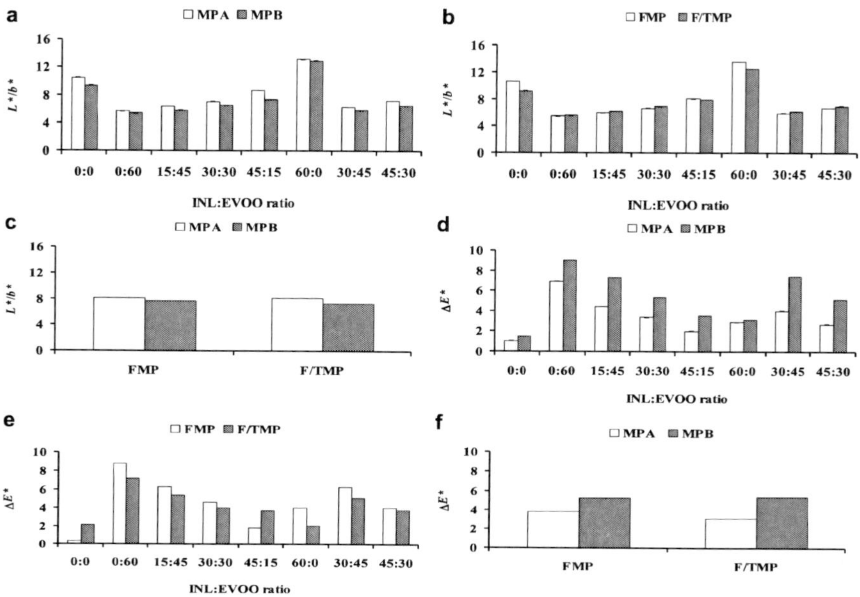

Figure 5. Instrumental colour measurements of MP samples: (a), (b), (c) white/yellow ratio; (d), (e), (f) total colour difference. MPA: mashed potatoes without added hydrocolloids; MPB: mashed potatoes with added hydrocolloids; FMP: fresh mashed potatoes; F/TMP: mashed potatoes subjected to a freeze/thaw cycle; INL: inulin; EVOO: extra virgin olive oil.

In turn, lack of colour in samples with added 60:0 ratio could be easily solved by addition of some β-carotene. Also, MPB samples had lower L^*/b^* values than MPA samples. In any case, colour parameter values of the MP samples with added 60:0 ratio indicate that the effect caused by hydrocolloid addition is weaker than the lightening produced by INL incorporation. By comparing both FMP and F/TMP samples (Figure 5b), it can be observed that at a fixed EVOO content, the increase in INL concentration did not significantly affect the colour of the MP samples (15:45 vs 30:45 and 30:30 vs 45:30); whereas at a fixed INL concentration addition of EVOO at higher concentrations considerably decreased L^*/b^* (30:30 vs 30:45 and 45:15 vs 45:30). In natural and commercial mashed potato, L^*/b^* values were lower than in the fresh controls after freezing and thawing of the samples, so that frozen-thawed were darker than fresh products [23]. Also, frozen mashed potato made from reconstituted potato flakes was darker than nonfrozen mashed potato [56].

In this study, note that the presence of EVOO significantly reduced differences in the L^*/b^* values among FMP samples and their F/TMP counterparts. As a consequence, a freeze/thaw cycle slightly decresased L^*/b^* ratio in the MPB samples but did not significantly affect it in the case of the MPA samples (Figure 5c). This is likely related to the water-holding capacity of the MPB samples, enhancing the overall light scattering of the MP.

Sample colour differences with respect to the fresh INL-EVOO-free control, based on the INL:EVOO ratio for both MPA and MPB samples (Figure 5d) and for both FMP and F/TMP samples (Figure 5e), corroborates that in both cases, the highest and lowest ΔE^* values were registered in the samples containing 0:60 and 60:0 ratios, respectively. As the droplet concentration increases, more reflected light travels through the oil phase of the MP being absorbed by the pigments mentioned earlier, intensifying the colour of the MP. Moreover, in the MPB samples, there were non-significant differences between the ΔE^* values of FMP and F/TMP samples (Figure 5f).

Quantitative Descriptive Analysis (QDA)

The Friedman test showed that there were significant differences between FMP samples due to INL:EVOO ratios and hydrocolloid addition (Tables 3, 4) with regards to *authentic colour*, *shine* and *uniformity* (χ^2 = 38.99, χ^2 = 73.96 and χ^2 = 68.51, respectively; P = 0.000); *consistency*, *adhesiveness*, *creaminess* and *fibrousness* (χ^2 = 104.72, χ^2 = 106.72, χ^2 = 76.94 and χ^2 = 66.52, respectively; P = 0.000) (Table 3) and *authentic taste*, *off-taste* and *ingredient taste* (χ^2 = 72.09, χ^2 = 106.10 and χ^2 = 107.44, respectively; P = 0.000) (Table 4). The panellists found that neither INL/EVOO combinations nor hydrocolloid addition had any detectable effect on *off-colours* or *sweetness* (χ^2 = 0.00 and χ^2 = 9.96, respectively) in the FMP products. On the other hand, although there were significant differences between FMP samples with respect to *uniformity* and *off-taste*, 60 and 90% confidence intervals were needed to pinpoint them respectively, so no *uniformity* differences between ranks due to INL/EVOO blends and hydrocolloid addition were established (Table 3).

F/TMP samples (Tables 5, 6) presented significant differences in terms of *authentic colour*, *shine* and *uniformity* (χ^2 = 91.89, χ^2 = 51.37 and χ^2 = 44.63, respectively; P = 0.000); *consistency*, *adhesiveness*, *creaminess* and *fibrousness* (χ^2 = 106.34, χ^2 = 106.10, χ^2 = 78.20 and χ^2 = 49.86, respectively; P = 0.000) (Table 5) and *sweetness*, *authentic taste*, *off-taste* and *ingredient taste* (χ^2 = 35.88, χ^2 = 45.66, χ^2 = 30 and χ^2 = 108.438, respectively; P = 0.000)

(Table 6). No INL:EVOO ratio or hydrocolloid addition effect on *off-colours* (χ^2 = 0.00) was detected by the panellists in the F/TMP products. However, although there were significant differences between F/TMP samples regarding *fibrousness*, *sweetness* and *authentic taste*, again a 90% confidence interval was needed to pinpoint them. In the case of F/TMP products, a 60% confidence interval was also needed to pinpoint the significant differences between F/TMP samples with respect to *uniformity* or *off-taste*, and therefore differences produced by INL/EVOO blends and hydrocolloid addition were not established for these attributes (Table 6).

Figures 6 and 7 show radar plots constructed to represent perceived sensory attributes. As far as sensory attributes describing appearance in FMP samples are concerned (Figure 6a), panellists gave FMPB-0:60, FMPA-15:45 and FMPB-30:45 samples higher scores for *authentic colour* than the rest of the samples, although only FMPA-0:0, FMPA-45:15, FMPA-60:0 and FMPB-60:0 differences were significant (Table 3), matching the results of instrumental colour measurements. Obviously, the colour of the ingredients used influences the colour of the product. INL is a white powder and when added alone produced a lighter-coloured mashed potato as indicated above. In INL-enriched dairy desserts, the average chain length of the added INL significantly affected brightness values but no other parameter. Samples with long-chain INL were the lightest in colour [57], corroborating these findings. Conversely, it has been found that INL did not affect the colour values of low-fat ice cream [58], and the addition of oligofructose to dairy drinks did not affect the L^*, a^* and b^* values [59]. In contrast, sensory analysis indicated that adding $\geq$ 45 g kg^{-1} of EVOO significantly improved colour due to the yellowness conferred by the oil. The panellists found that increasing yellowness is desirable since, as mentioned above, the flesh of Kennebec potatoes is white and these MP without other added ingredients look rather pale. In the case of *shine*, the panellists found a significant difference between the FMPA-0:60 sample, which scored highest, and FMPA-0:0 and FMPB-0:0 controls, neither of which contained any EVOO (Figure 6a). Note that in the case of *uniformity* the panellists also identified FMPA-0:0 and FMPA-60:0 samples as having lower scores (Table 3) than all the other FMP samples. This could be ascribed to a lack of hydrocolloid-water interactions in the absence of XG, as well as to the absence of oil droplets, detected by the panellists.

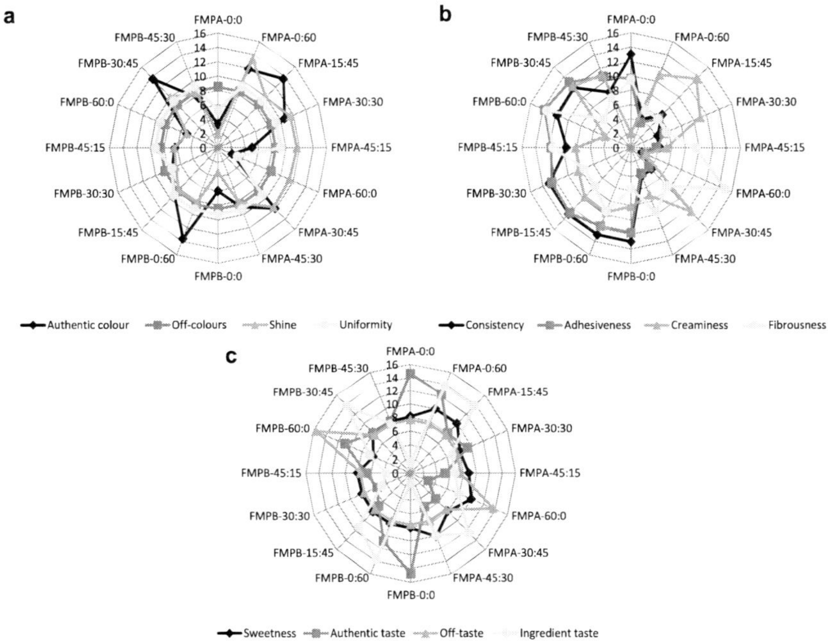

Figure 6. Sensory attributes as functions of INL:EVOO ratio in FMP samples: (a) appearance sensory attributes; (b) texture sensory attributes; (c) taste sensory attributes. Identification of system notation in Table 1. INL: inulin; EVOO: extra virgin olive oil.

As can be seen from the sensory appearance attributes (Figure 7a) of the F/TMP samples, F/TMPA-60:0 had the lowest score for *authentic colour* (Table 5)and sample F/TMPB-45:15 scored the highest, though differences with F/TMPA-0:60, F/TMPB-0:60, F/TMPA-15:45, F/TMPB-15:45, F/TMPA-30:30, F/TMPB-30:30, F/TMPA-45:15, F/TMPA-30:45, F/TMPB-30:45, F/TMPA-45:30 and F/TMPB-45:30 samples were nonsignificant. It is obvious that the presence of EVOO, even at the lowest level, masked the lightening effect caused by INL incorporation, suggesting that panellists only detected loss of colour when samples with added INL alone were subjected to a freeze/thaw cycle. In the case of *shine*, the panellists distinguished a marked difference between the F/TMPB-0:0 control, which obtained the lowest score, and the F/TMPA-0:60 and F/TMPB-30:45 samples with the highest scores (although differences between F/TMPA-0:60 and F/TMPB-30:45 samples were nonsignificant).

Table 3. Effects of INL:EVOO ratio and hydrocolloid addition on mean ranks of sensory appearance and texture attributes for FMP samples

Sample code	Authentic colour	Off-colours	Shine	Uniformity	Consistency	Adhesiveness	Creaminess	Fibrousness
FMPA-0:0	$3.25*_{c}$	8.50	1.88_{c}	5.25*	13.00_{a}	$9.62*_{abcd}$	$2.00*_{c}$	$9.50*_{ab}$
FMPB-0:0	6.00_{abc}	8.50	3.38_{b}	9.25	13.00_{a}	11.75_{abc}	8.12_{abc}	5.62_{b}
FMPA-0:60	11.88_{ab}	8.50	$13.38*_{a}$	9.25	$4.25*_{bc}$	$3.75*_{cd}$	11.00_{ab}	$5.62*_{b}$
FMPB-0:60	13.62_{a}	8.50	8.25_{abc}	9.25	13.00_{a}	11.75_{abc}	9.62_{abc}	9.50_{ab}
FMPA-15:45	$13.62*_{a}$	8.50	10.00_{abc}	9.25	$6.50*_{abc}$	$5.75*_{abcd}$	$13.62*_{a}$	5.62_{b}
FMPB-15:45	9.12_{abc}	8.50	8.25_{abc}	9.25	13.00_{a}	12.75_{ab}	9.62_{abc}	7.50_{ab}
FMPA-30:30	$10.62*_{abc}$	8.50	11.50_{ab}	9.25	$4.25*_{bc}$	$5.25*_{bcd}$	$11.00*_{ab}$	5.62_{b}
FMPB-30:30	7.38_{abc}	8.50	6.75_{abc}	7.38	13.00_{a}	12.62_{ab}	8.38_{abc}	7.50_{ab}
FMPA-45:15	5.00_{bc}	8.50	$11.62*_{ab}$	9.25	$4.25*_{bc}$	$3.75*_{cd}$	$5.25*_{abc}$	9.50_{ab}
FMPB-45:15	6.38_{abc}	8.50	6.50_{abc}	9.25	9.50_{abc}	11.75_{abc}	8.12_{abc}	11.50_{ab}
FMPA-60:0	$2.25*_{c}$	8.50	$11.38*_{ab}$	3.12*	$1.75*_{c}$	$2.25*_{d}$	$2.38*_{c}$	14.88_{a}
FMPB-60:0	5.00_{bc}	8.50	5.00_{abc}	9.25	11.75_{ab}	13.75_{a}	4.12_{bc}	13.38_{ab}
FMPA-30:45	11.88_{ab}	8.50	11.62_{ab}	9.25	$4.25*_{bc}$	$3.75*_{cd}$	12.50_{ab}	7.50_{ab}
FMPB-30:45	13.62_{a}	8.50	10.00_{abc}	9.25	11.88_{ab}	12.88_{ab}	12.12_{ab}	5.62_{b}
FMPA-45:30	8.88_{abc}	8.50	8.25_{abc}	9.25	$4.25*_{bc}$	$3.88*_{cd}$	$7.12*_{abc}$	$9.62*_{ab}$
FMPB-45:30	7.50_{abc}	8.50	8.25_{abc}	9.25	8.38_{abc}	10.75_{abc}	11.00_{ab}	7.50_{ab}
χ^2 value	38.99	0.00	73.96	68.51	104.72	106.72	76.94	66.52
P value	0.000	1.000	0.000	0.000	0.000	0.000	0.000	0.000

a-d For each sensory attribute mean ranks without the same letter are significantly different ($P < 0.05$). *For each sensory attribute and for the same.

INL: EVOO ratio mean ranks between FMPA samples and their FMPB counterparts are significantly different ($P < 0.05$); FMPA: fresh mashed potatoes without added hydrocolloids; FMPB: fresh mashed potatoes with added hydrocolloids. INL, inulin; EVOO, extra virgin olive oil.

Table 4. Effects of INL:EVOO ratio and hydrocolloid addition on mean ranks of sensory taste attributes for FMP samples

Sample code	Sweetness	Authentic taste	Off-taste†	Ingredient taste
FMPA-0:0	8.12	14.62_{a}	7.62_{b}	1.50_{d}
FMPB-0:0	8.12	14.75_{a}	7.62_{b}	1.50_{d}
FMPA-0:60	10.00	12.38_{ab}	7.62_{b}	13.75_{a}
FMPB-0:60	8.12	10.88_{abc}	7.62_{b}	13.75_{a}
FMPA-15:45	10.00	8.00_{abc}	7.62_{b}	13.75^{*}_{a}
FMPB-15:45	8.12	6.88_{abc}	7.62_{b}	11.25_{abc}
FMPA-30:30	8.12	9.38^{*}_{abc}	7.62_{b}	6.75_{abcd}
FMPB-30:30	8.00	5.52_{abc}	7.62_{b}	4.75_{bcd}
FMPA-45:15	8.88	5.25_{bc}	7.62_{b}	6.75^{*}_{abcd}
FMPB-45:15	8.12	6.75_{abc}	7.62_{b}	3.75_{cd}
FMPA-60:0	10.00*	2.88^{*}_{c}	13.62_{ab}	7.88_{abcd}
FMPB-60:0	6.12	10.88_{abc}	15.62_{a}	6.75_{abcd}
FMPA-30:45	8.00	5.38^{*}_{bc}	7.62_{b}	12.50_{ab}
FMPB-30:45	8.12	8.00_{abc}	7.62_{b}	13.75_{a}
FMPA-45:30	10.12	5.38^{*}_{bc}	7.62_{b}	10.00^{*}_{abc}
FMPB-45:30	8.00	8.00_{abc}	7.62_{b}	7.62_{abcd}
χ^2 value	9.96	72.09	106.10	107.44
P value	0.822	0.000	0.000	0.000

a-d For each sensory attribute mean ranks without the same letter are significantly different ($P < 0.05$). † For *off-taste* mean ranks without the same letter are significantly different ($P < 0.10$). * For each sensory attribute and for the same INL:EVOO ratio mean ranks between FMPA samples and their FMPB counterparts are significantly different ($P < 0.05$); FMPA: fresh mashed potatoes without added hydrocolloids; FMPB: fresh mashed potatoes with added hydrocolloids; INL, inulin; EVOO, extra virgin olive oil.

With regards to the sensory texture attributes of FMP samples (Figure 6b, Table 3), in the case of *consistency* the panellists gave significantly higher scores to samples FMPA-0:0, FMPB-0:0, FMPB-0:60, FMPB-15:45 and FMPB-30:30 than to all the other samples (except for FMPA-15:45, FMPB-45:15, FMPB-60:0, FMPB-30:45 and FMPB45:30 samples). FMPA-60:0 scored lowest for this texture attribute. Therefore, the sensory panel indicated that 60 g kg^{-1} INL provided superior softening in FMP samples than 60 g kg^{-1} EVOO. Note that except for both controls, all the FMPB samples had higher scores for *consistency* than their FMPA counterparts. Conversely, both FMPA and F/TMPA samples were more elastic, with a significantly higher magnitude of G' than their counterparts with added hydrocolloids. A previous study showed that in MP samples, κ-C provided the appropriate texture, while XG imparted creaminess to the product, which was associated with an increase in the number of XG-water interactions [21]. The effect of κ-C on the thickening of the product was ascribed to the interaction of κ-C with denatured milk protein, while XG exhibited a considerably greater capacity (even at very low concentration) to retain water than κ-C. Therefore, it would appear that panellists were more sensitive to the effect of carrageenan in the products with added hydrocolloids, whereas rheological measurements would seem to measure the water holding capacity conferred by the gum. In the case of *adhesiveness* (Table 3), the panellists distinguished significant differences between the FMPB-60:0 sample, which scored the highest, and FMPA-0:60, FMPA-30:30, FMPA-45:15, FMPA-60:0, FMPA-30:45 and FMPA-45:30 samples, which obtained lower scores, possible because they did not contain hydrocolloids, thus confirming previous findings [16, 19, 21]. This fact is mainly ascribed to the absence of κ-C, indicating that *adhesiveness* is also affected by the gelling properties of the carrageenan.

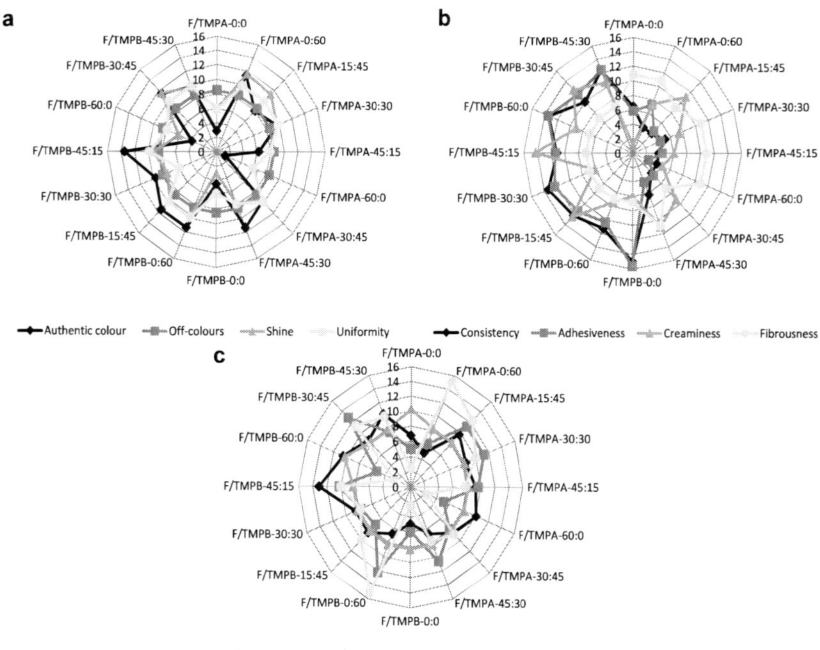

Figure 7. Sensory attributes as functions of INL:EVOO ratio in F/TMP samples: (a) appearance sensory attributes; (b) texture sensory attributes; (c) taste sensory attributes. Identification of system notation in Table 1. INL: inulin; EVOO: extra virgin olive oil.

For *creaminess* (Figure 6b), panellists distinguished a significant difference between the FMPA-0:0 control and the FMPA-60:0 sample with added INL alone, both of which had a lower score, and the FMPA-0:60, FMPA-15:45, FMPA-30:30, FMPA-30:45, FMPB-30:45 and FMPB-45:30 samples, which presented a softer mouthfeel. Therefore, by adding EVOO to FMP samples at concentrations ≥ 30 g kg^{-1}, either individually or blended with INL, increased their *creaminess* compared with the control without added ingredients or the product with added INL alone. In the experiments referred to in this chapter, panellists gave the FMPA-60:0 sample the highest score for *fibrousness*, although only the differences between this score and those for FMPB-0:0, FMPA-0:60, FMPA-15:45, FMPA-30:30 and FMPB-30:45 were significant (Table 3). An extensive description of creaminess, based on discussions with consumers, includes a range of sensations typically associated with fat content, such as a full sweet taste, compact smoothness, not rough, not dry, with a velvety (not oily) coating [11]. In various studies it has been shown that creaminess is strongly related to other sensory characteristics, such as the $\eta_{app,50}$ which was a principal factor influencing creaminess in MP with added XG, mainly at concentrations up to 1.5 g kg^{-1} [17]. For only dairy products a combination of thickness and smoothness was sufficient to predict creaminess reasonably well [60]. However, for other products creaminess was less well explained [61] indicating that other factors are also important, like for instance, fat content [62, 63]. Creaminess is negatively affected by texture attributes such as roughness [64, 65], and heterogeneous mouthfeel [64, 66]. On the other hand, in dairy desserts, the addition of long-chain INL gave rise to a low-fat product with the same thickness intensity as the full-fat sample [45], though less creamy, rougher and with more mouth coating, which was attributed by the latter authors to aggregate formation characterised by a rough outline.

In the F/TMP samples (Figure 7b, Table 5), panellists scored the F/TMPB-0:0 control higher for *consistency* and *adhesiveness* than any of the other samples, matching the results of oscillatory properties (Figure 2d). Furthermore, differences between F/TMPB-0:0 and the rest of the frozen/thawed samples with added hydrocolloids (F/TMPB samples) were nonsignificant for both *consistency* and *adhesiveness*, irrespective of the INL:EVOO ratio. A previous study showed that INL concentrations were responsible for minor difference in most rheological properties, firmness and overall acceptability of MP samples, whereas the addition of κ-C and XG resulted in the main differences between samples [26]. As the effect of INL on the thickening of the product was limited, the addition of small amounts of κ-C and XG to MP samples significantly improved freeze/thaw stability, confirming previous findings [21]. On the other hand, this result confirms that in the case of olive oil, a liquid product, textural attributes are not as significant as those of aroma and taste [9]. Panellists scored sample F/TMPB-45:15 the highest for *creaminess*, although only the differences between the score for this sample, and those for F/TMPA-0:0 control and F/TMPA-60:0, which had the highest INL concentrations, were significant (Table 5). Long-chain INL is used as a fat replacer in several dairy products [45] due to its ability to act as a fat mimetic or fat replacer, based on its capacity to form microcrystals, which interact with each other, forming small aggregates that occlude a great amount of water and that may ultimately agglomerate into a gel network [4, 49, 55]. However, the fat mimetic capacity of INL depends not only on the chain length, but also on the concentration of added INL [67]. Long-chain INL has been used in low-fat yogurts as a fat replacer where it was shown to significantly improve creaminess, mouthfeel, and smoothness [68, 69]. The addition of long-chain INL also improved creaminess and consistency of low-fat custards [70] and mimicked those of full-fat custard [71]. Probably, INL at 45 g kg^{-1} enhanced creaminess in the F/TMPB samples thanks to the lubricating and coating properties either of the gum or the oil. In the case of *fibrousness* (Figure 7b, Table 5), the most important point to note is that panellists perceived greater *fibrousness* in the F/TMPA samples than in all their F/TMPB counterparts. Therefore, it was obvious that the assessors associated decreased *fibrousness* with the presence of hydrocolloids retarding starch retrogradation [37].

As regards the sensory taste attributes (Figure 6c, Table 4) of the FMP samples, in the case of *authentic taste*, the panellists noted that the FMPA-0:0 and FMPB-0:0 controls, which had the highest scores, differed significantly from the FMPA-45:15, FMPA-60:0, FMPA-30:45 and FMPA-45:30 samples, which scored the lowest. Panellists considered that MP without added ingredients produced an unpleasant taste in the mouth due to an excessively strong potato flavour. In addition, panellists scored the sample with added hydrocolloids and added INL alone, i.e., FMPB-60:0, significantly higher for *off-taste* than the rest of the samples. Analogously, increasing INL concentration by up to 6% significantly reduced the overall acceptability of all the FMPA, FMPB, F/TMPA and F/TMPB samples [26]. Panellists found that the addition of 6% INL gave the F/TMPB samples an unfamiliar taste and odour, despite the fact that differences between scores for both F/TMPA and F/TMPB samples were nonsignificant. Although INL has been considered to have a bland neutral taste without any off-flavour or aftertaste [72], its effect on the flavour of vanilla dairy desserts may be different from the effect of fat.

The obvious influence of INL on sweetness and vanilla flavour intensity in skimmed milk samples with lower starch concentrations suggests a faster flavour release in these systems [71]. Conversely, samples FMPA-0:60, FMPB-0:60, FMPA-15:45 and FMPB-30:45 scored highest for *ingredient taste* (Table 4). Panellists ascribed these differences to an aromatic note detected in the MP samples with added EVOO at the highest concentrations. EVOO must have an acidity below 0.8% (like oleic acid) and a flavour that results purely from the fruit's metabolism, which yields fruity, bitter, and pungent characteristics, with no defects (a fusty, musty, winey, or rancid taste) [9]. Volatile compounds stimulate the olfactory receptors causing the perception of the whole aroma of the virgin olive oil [73]. The sensory characteristics of 22 EVOOs from California, Italy, Spain, Chile, and Australia were compared using a generic descriptive analysis [9]. With the exception of "thick" and "citrus", all sensory attributes were significantly different among the oils. Spanish EVOOs were characterised by their tropical fruit, herb, mint, floral, bitter, and pungent attributes. The effects of fat on odour and flavour attributes may be related to the flavour-releasing properties of fat. In sixteen virgin olive oils, volatile and phenolic compounds have been related to olfactory and gustative properties respectively [74]. The volatile composition revealed several correlations with olfactory attributes perceived by sensory analysis: the sum of aldehydes C_6 was correlated with the orthosonal perception of fruity olive and the retrosanal odour of almond.

F/TMPB-45:15 scored highest among the F/TMP products (Figure 7c, Table 6), as regards *sweetness*, whereas F/TMPB-0:0 and F/TMPA-0:60 samples had the lowest scores for this taste attribute. This indicates that, as expected, INL addition increased the sugariness of the F/TMP samples because of its higher total soluble solid content (TSS). In both MPA and MPB samples, the highest TSS values corresponded to samples with added INL alone (60:0 ratio) and were higher in the MPB samples, which is logical as two polysaccharides were also added to these MP samples [42]. Supplementation with INL introduced more reducing sugars and led to increased total carbohydrate content [75].

Similar results were obtained by Castro et al. [59] in dairy drinks with added olifructose and by González-Tomás et al. [57] and Villegas et al. [67] in dairy drinks with added INL. Panellists scored F/TMPB-0:60, F/TMPA-15:45, F/TMPA-30:30, F/TMPA-45:15, F/TMPB-45:15, F/TMPB-30:45 and F/TMPA-45:30 samples higher for *authentic taste* than the F/TMPA-0:0 and F/TMPB-0:0 controls (Table 6). A freeze/thaw cycle clearly mitigated the strong potato flavour detected in the fresh controls (Table 4). Panellists gave F/TMPA-0:60 and F/TMPB-0:60 samples the highest score for *ingredient taste* (Figure 7c), although only the differences between scores for these samples and those for samples without added EVOO (F/TMPA-0:0, F/TMPB-0:0, F/TMPA-60:0 and F/TMPB-60:0 samples) were significant. This means that a freeze/thaw cycle did not affect the aroma and taste attributes of EVOO.

Table 5. Effects of INL:EVOO ratio and hydrocolloid addition on mean ranks of sensory appearance and texture attributes for F/TMP samples

Sample code	Authentic colour	Off-colours	Shine	Uniformity	Consistency	Adhesiveness	Creaminess	Fibrousness†
F/TMPA-0:0	2.88_{c}	8.50	$6.00*_{ab}$	5.88	$6.50*_{bc}$	$5.88*_{bcd}$	$1.12*_{c}$	$10.88*_{a}$
F/TMPB-0:0	4.50_{bc}	8.50	3.00_{b}	5.88	15.00_{a}	15.62_{a}	6.00_{abc}	6.88_{b}
F/TMPA-0:60	11.38_{ab}	8.50	11.62_{a}	9.88	$3.88*_{c}$	$7.25*_{abcd}$	7.38_{abc}	$10.88*_{a}$
F/TMPB-0:60	11.38_{ab}	8.50	9.62_{ab}	9.88	11.25_{abc}	10.38_{abcd}	7.38_{abc}	6.88_{b}
F/TMPA-15:45	$8.00*_{abc}$	8.50	11.25_{ab}	9.88	$3.88*_{c}$	$4.38*_{cd}$	11.00_{ab}	8.88_{ab}
F/TMPB-15:45	11.38_{ab}	8.50	9.62_{ab}	9.88	12.25_{abc}	11.38_{abc}	12.50_{ab}	6.88_{b}
F/TMPA-30:30	9.62_{abc}	8.50	9.62_{ab}	9.88*	$5.12*_{bc}$	$4.38*_{cd}$	7.38_{abc}	$10.88*_{a}$
F/TMPB-30:30	9.62_{abc}	8.50	7.75_{ab}	5.88	13.38_{ab}	12.38_{abc}	6.00_{abc}	6.88_{b}
F/TMPA-45:15	$6.25*_{abc}$	8.50	7.75_{ab}	7.88	$3.88*_{c}$	$4.38*_{cd}$	$6.12*_{abc}$	$10.88*_{a}$
F/TMPB-45:15	13.50_{a}	8.50	9.25_{ab}	9.88	11.12_{abc}	11.38_{abc}	14.25_{a}	6.88_{b}
F/TMPA-60:0	$1.38*_{c}$	8.50	6.00_{ab}	5.88	$3.88*_{c}$	$2.50*_{d}$	$4.75*_{bc}$	$10.88*_{a}$
F/TMPB-60:0	3.88_{bc}	8.50	6.00_{ab}	7.88	13.38_{ab}	13.50_{ab}	9.00_{abc}	6.88_{b}
F/TMPA-30:45	9.75_{abc}	8.50	9.62_{ab}	9.88	$3.88*_{c}$	$4.38*_{cd}$	$9.12*_{abc}$	6.88_{b}
F/TMPB-30:45	11.38_{ab}	8.50	11.50_{a}	9.88	10.00_{abc}	11.50_{abc}	12.38_{ab}	6.88_{b}
F/TMPA-45:30	11.38_{ab}	8.50	7.75_{ab}	7.88	$6.25*_{bc}$	$4.38*_{cd}$	11.00_{ab}	$10.88*_{a}$
F/TMPB-45:30	9.75_{abc}	8.50	9.62_{ab}	9.88	12.38_{ab}	12.38_{abc}	10.62_{ab}	6.88_{b}
χ^2 value	91.89	0.00	51.37	44.63	106.34	106.10 7	78.20	49.86
P value	0.000	1.000	0.000	0.000	0.000	0.000	0.000	0.000

a-d For each sensory attribute mean ranks without the same letter are significantly different ($P < 0.05$). † For *fibrousness* mean ranks without the same letter are significantly different ($P < 0.10$). * For each sensory attribute and for the same INL:EVOO ratio mean ranks between F/TMPA samples and their F/TMPB counterparts are significantly different ($P < 0.05$); F/TMPA: frozen/thawed mashed potatoes without hydrocolloids; F/TMPB: frozen/thawed mashed potatoes with added hydrocolloids. INL, inulin; EVOO: extra virgin olive oil.

Table 6. Effects of INL:EVOO ratio and hydrocolloid addition on mean ranks of sensory taste attributes for F/TMP samples

Sample code	Sweetness†	Authentic taste†	Off-taste	Ingredient taste
F/TMPA-0:0	6.75_{ab}	5.00_{e}	10.25	2.50_{b}
F/TMPB-0:0	4.88_{b}	6.00_{de}	8.25	2.50_{b}
F/TMPA-0:60	4.88_{b}	$6.12*_{de}$	8.25	15.12_{a}
F/TMPB-0:60	6.75_{ab}	12.25_{a}	8.25	15.12_{a}
F/TMPA-15:45	9.75_{ab}	$11.25*_{ab}$	8.25	$12.38*_{a}$
F/TMPB-15:45	8.62_{ab}	7.12_{cde}	8.25	10.00_{ab}
F/TMPA-30:30	8.50_{ab}	$11.25*_{ab}$	8.25	$10.00*_{ab}$
F/TMPB-30:30	8.50_{ab}	7.75_{bcde}	8.25	7.75_{ab}
F/TMPA-45:15	$9.25*_{ab}$	9.62_{abcd}	8.25	$7.75*_{ab}$
F/TMPB-45:15	13.12_{a}	10.25_{abc}	8.25	10.00_{ab}
F/TMPA-60:0	10.12_{ab}	5.12_{e}	8.25	2.50_{b}
F/TMPB-60:0	10.50_{ab}	5.25_{e}	10.25	2.50_{b}
F/TMPA-30:45	8.62_{ab}	$7.88*_{bcde}$	8.25	$8.88*_{ab}$
F/TMPB-30:45	8.62_{ab}	12.75_{a}	8.25	11.25_{a}
F/TMPA-45:30	$6.75*_{ab}$	$10.62*_{abc}$	8.25	$7.75*_{ab}$
F/TMPB-45:30	10.38_{ab}	7.75_{bcde}	8.25	10.00_{ab}
χ^2 value	35.88	45.66	30.00	108.38
P value	0.002	0.000	0.012	0.000

a-d For each sensory attribute mean ranks without the same letter are significantly different ($P < 0.05$).

† For *sweetness* and *authentic taste* mean ranks without the same letter are significantly different ($P < 0.10$).

* For each sensory attribute and for the same INL:EVOO ratio mean ranks between F/TMPA sample and its F/TMPB counterpart are significantly different ($P < 0.05$); F/TMPA: frozen/thawed mashed potatoes without added hydrocolloids; F/TMPB: frozen/thawed mashed potatoes with added hydrocolloids. INL, inulin; EVOO, extra virgin olive oil.

Linear Discriminat Analysis (LDA)

To sum up, the Friedman test confirms significant variability among the sensory attributes of MP samples under the different experimental conditions. The purpose of the application of LDA analyses was to focus on the possible objective discrimination between the different ingredients added to MP samples on the basis of the sensory attributes considered in the QDA. LDA analyses were designed to develop a small number of functions with which to discriminate among the different performance conditions of MP based on linear combinations of eleven sensory attributes. In the LDA analyses, one colour attribute (*off-colours*) was rejected for the model as its variance was zero.

Table 7 summarises the generation of the discriminant functions for the four different LDA analyses performed on the FMP and F/TMP samples considered separately (n = 128), and using hydrocolloid addition and INL:EVOO ratio as discriminating factors respectively. Only significant discriminating functions are shown. The relative percentages of variance accounting for discriminating function 1 were clearly higher in all the analyses. For LDAs using hydrocolloid addition as discriminating factor, note that function 1 is good alone, with 100% relative percentages in both FMP and F/TMP samples.

Table 7. Summary of the discriminant analysis results for each discriminating factor used in FMP and F/TMP samples

Discriminant function	Eigenvalue	Relative percentage	Canonical correlation	*P* value
Discriminating factor: hydrocolloid addition				
FMP samples				
1	3.50	100	0.88	0.000
F/TMP samples				
1	6.26	100	0.93	0.000
Discriminating factor: INL:EVOO ratio				
FMP samples				
1	14.83	67.4	0.97	0.000
2	5.50	25.0	0.92	0.000
3	1.18	5.4	0.74	0.000
4	0.31	1.4	0.48	0.017
F/TMP samples				
1	18.53	90.90	0.97	0.000
2	1.19	5.80	0.74	0.000
3	0.35	1.70	0.51	0.011

FMP: fresh mashed potatoes; F/TMP: frozen/thawed mashed potatoes; INL, inulin; EVOO, extra virgin olive oil.

In turn, when considering INL:EVOO ratio as discriminating factor, function 1 had the highest relative percentage (> 90%) for the F/TMP samples, and canonical correlation of 0.97. This evidences that detection of the effect of the INL:EVOO ratio by the panellists was more significant in the samples that had been subjected to a freeze/thaw cycle. As shown above, mechanical spectra confirmed some structural changes occurring in the MP samples after a freeze/thaw cycle as a consequence of amylopectin retrogradation process. In addition, it has been reported that in F/TMPA samples with added oil the structural damage caused by freezing enabled the oil droplets to come close enough together to aggregate, while in the F/TMPB ones with added κ-C and XG and oil, the XG is adsorbed onto the surface of the droplets, enhancing stability against flocculation and coalescence and forming a white film as observed by scanning electron microscopy [76].

Table 8. Standardised discriminant function coefficients for the four LDA analyses carried out in the FMP and F/TMP samples using different discriminat factors

Independent variables	Discriminant function			
	1	2	1	2
Discriminating factor: hydrocolloid addition				
	FMP samples		F/TMP samples	
Authentic colour	-0.29	-	-0.25	-
Shine	0.31	-	0.14	-
Uniformity	0.18	-	-0.07	-
Consistency	0.39	-	**0.97**	-
Adhesiveness	**0.98**	-	0.48	-
Creaminess	**0.41**	-	0.23	-
Fibrousness	0.07	-	-0.16	-
Sweetness	0.10	-	0.28	-
Authentic taste	-0.33	-	-0.08	-
Off-taste	0.35	-	0.19	-
Ingredient taste	-0.04	-	**0.65**	-
Discriminating factor: INL:EVOO ratio				
	FMP samples		F/TMP samples	
Authentic colour	0.15	-0.29	0.16	-0.06
Shine	-0.21	-0.11	-0.18	0.51
Uniformity	-0.27	0.38	**0.28**	0.16
Consistency	0.25	-0.21	-0.01	0.25
Adhesiveness	-0.04	0.02	-0.12	**-0.90**
Creaminess	0.63	-0.14	0.12	**1.05**
Fibrousness	**0.74**	**0.51**	0.19	0.07
Sweetness	0.16	0.09	0.03	0.30
Authentic taste	0.05	-0.10	-0.03	0.28
Off-taste	0.35	**0.95**	-0.21	0.01
Ingredient taste	**1.25**	-0.03	**1.00**	-0.32

FMP: fresh mashed potatoes; F/TMP: frozen/thawed mashed potatoes; INL, inulin; EVOO, extra virgin olive oil.

In general oil retention from an emulsion was found to increase with decreasing stability of the emulsions towards coalescence, either due to lowering the adsorb amount of protein at the oil water interface [77], adapting the composition of the adsorbed layer [78] or by partial coalescence [79]. For reversible aggregating emulsions this lead to increased sensations of fat-related texture attributes. It resulted in higher ratings for a group of viscosity-related attributes as thickness, creaminess, and fatiness and lower ratings for a group of friction related attributes as roughness, dryness and astringency [78, 79]. Therefore, a higher degree of coalescence before intake could justify that the panellists had increased sensory perception of the INL:EVOO ratio in the samples subjected to a freeze/thaw cycle.

Table 8 shows the standardised coefficients of the functions used to discriminate among the different levels of the discriminating factor used in the LDA. The relative magnitude of the coefficients in this table shows how the sensory attributes are used to discriminate among the groups. In the case of the only discriminant function 1 extracted by the LDA in which the addition of hydrocolloids is the discriminating factor for the FMP samples, *adhesiveness* is the sensory attribute that best separates samples without and with added hydrocolloids (FMPA and FMPB samples, respectively). Also, in the case of discriminant function 1 where hydrocolloid addition is the discriminating factor for the F/TMP samples, first *consistency* and then *ingredient taste* are the variables that best separate the samples without and with added hydrocolloids (F/TMPA and F/TMPB samples, respectively).

In the case of the LDA using INL:EVOO ratio as a discriminating factor in the FMP samples, for the first discriminant function *ingredient taste* is clearly the sensory attribute that best separates the eight levels of INL/EVOO blends added to MP, whereas for the second function first *off-taste* and then *fibrousness* are the attributes that best separate the levels of INL:EVOO ratio. Also in the case of discriminant function 1 (more significant) extracted by the LDA using the INL:EVOO ratio as a discriminating factor, again *ingredient taste* is the sensory attribute which best separate the eight INL:EVOO ratios in the F/TMP samples, while for the second function first *creaminess* and then *adhesiveness* are the attributes that best separate the levels of of INL/EVOO blends.

Table 9. Group centroids for the LDA analysis carried out in FMP and F/TMP samples using INL:EVOO ratio as discriminant factor

	FMP samples		F/TMP samples	
Group centroids	**1**	**2**	**1**	**2**
INL:EVOO ratio				
0:0	-8.09	-0.50	-7.08	-1.74
0:60	3.70	-1.44	4.86	-1.44
15:45	3.02	-1.33	2.81	0.73
30:30	-1.68	-1.06	1.50	0.61
45:15	-2.47	0.37	1.22	-0.63
60:0	1.65	5.79	-6.86	1.39
30:45	3.31	-1.48	2.07	0.61
45:30	0.56	-0.35	1.48	0.48

FMP: fresh mashed potatoes; F/TMP: frozen/thawed mashed potatoes; INL, inulin; EVOO, extra virgin olive oil.

Table 9 shows the group centriods and Figure 8 shows the projections of the observations onto the two axes extracted by the LDA using the INL:EVOO ratio as a discriminating factor in the FMP and F/TMP samples each considered separately. As expected from the high relative percentages of the discriminant functions 1, mainly in the F/TMP samples, the target effect is better defined by the function 1, while there is more overlap in the case of the function 2.

In the case of LDA using FMP samples (Figure 8a), centroid values show that the first function discriminates mainly between 0:0 control and the rest of the samples, (it separates them by more than 6 units; Table 9). This result is clearly due to that the panellists scored both FMPA and FMPB samples with added 0:0 ratio higher for *autenthic taste* (Figure 6c; Table 4) due to the strong potato taste detected in the samples without other ingredients added. The FMPA samples with a 0:0 addition ratio also scored quite high for *fibrousness*—in this case influenced by the low *creaminess* detected in the samples without either κ-C and XG or other ingredients added. The most obvious discrimination was between 0:0 control and samples containing 60 g kg^{-1} added EVOO alone and those containing 45 g kg^{-1} added EVOO and 15 and 30 g kg^{-1} added INL respectively, which received the highest *ingredient taste* scores (Table 4) due to their high oil content and the lowest *fibrousness* scores (Table 3). Note that discrimination between 0:60, 15:45, and 30:45 ratios was quite poor (Figure 8a), partially due to the similar *ingredient taste* scores given by the panellists to the samples containing these three

INL/EVOO blends. It turned out to be insufficient to know in detail the chemical composition of the product and the threshold values of these components for taste and odour perception [11]. Both the bitter taste and the trigeminal sensation of pungency stimulate the taste receptors and the trigeminal nerve [9]; those sensory properties have been associated with phenolic compounds present in EVOO [7]. Also, the structure of the food product appeared to be a main factor in determining taste and odour perception [80]. High fat custards produced less sensation of dryness and roughness, more sensations of flavour, and more sensations of creamy and fatty mouth and after feel than their zero-fat containing counterparts [81]. Addition of extra virgin olive oil in increasing concentrations enhanced the sensory quality of MP in terms of reduced granularity, denseness, cohesiveness, adhesiveness and fibrousness, and increased homogeneity, ease of swallowing and palate coating [76]. In turn, presence of EVOO in MP significantly increased the overall acceptability of the product which was ascribed by the panellists chiefly to lubricating and coating properties conferred by the oil [42]. It is clear that many sensory attributes cannot be related to a single physical property in a single way [11].

For the FMP samples, centroid values show that the second function only discriminates between samples with 60 g kg^{-1} added INL alone and the rest (Table 9, Figure 8a). This ties up with the fact that the addition of INL at 60 g kg^{-1} produced an unpleasant and unfamiliar taste as evidenced by the higher *off-taste* scores (Figure 6c, Table 4) awarded to these samples. In addition, this result is also due to that the panellists scored the FMP samples with added 60:0 ratio higher for *fibrousness* (Table 3) due to the sandy texture detected in the samples with higher INL contents. Concerning this observation, González-Tomás et al. [57] investigated how adding INL of different average chain lengths (long-chain, native, and short-chain INL) at a concentration of 7.5% (wt/wt) would affect the physicochemical and sensory characteristics of starch-based dairy desserts formulated with either skim or whole milk. The results showed that the effect of adding 7.5% INL of different average chain length can give rise to products with different rheological behaviour and different sensory characteristics. The skim milk sample with long-chain INL and the whole milk sample without INL were perceived to have the same creaminess and consistency intensity, but addition of long-chain INL also increased roughness intensity and, consequently, the sensory quality was negatively affected. This effect could be due to small crystals or crystal aggregates of long-chain INL in the product that influence the sensory characteristics of the samples [42]. Some authors have reported on the particle

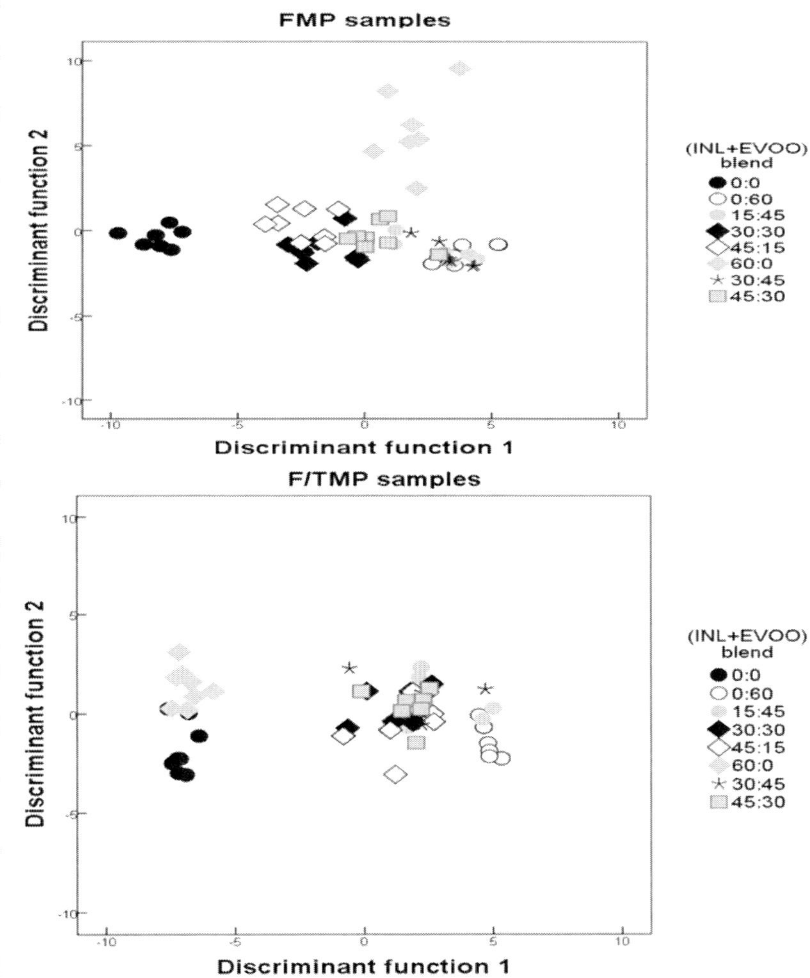

Figure 8. Plot of observations on the two principal axes obtained via LDA based on the INL:EVOO ratio used as discriminating factor for: (a) FMP samples; (b) F/TMP samples; INL: inulin; EVOO: extra virgin olive oil.

size of INL crystals [45], whereas others have discussed the effects that adding different sized particles may have on the sensory characteristics of dairy matrices. Hebette et al. [50] described the complex melting and crystalline behaviour of an aqueous suspension of INL crystallites. They observed the presence of 8-like shaped INL crystallites with particle size from 5 to 20 μm by optical microscopy. Kim et al. [82] also found differences in INL particle sizes between low shearing, high shearing, and thermally induced gels. The average particle size of INL crystals varied from 30 μm for low-shear induced gel to approximately 2 μm for thermally induced gels. Therefore, INL crystals with a particle size from 2 to 30 μm could be expected in the FMP samples. According to this information, the increase in *fibrousness* perceived for FMP samples with added long-chain INL alone could be explained by the presence of INL crystals, confirming previous findings [26, 42].

In the case of F/TMP samples (Figure 8b), function 1 discriminates clearly between 0:0 control and samples with 60 g kg^{-1} added I alone, and the rest. Again, this is chiefly due to the lower *ingredient taste* scores given by the panellists to the samples without added EVOO. Centroid values (Table 9) show that function 1 discriminates less between samples with addition ratios of 15:45, 30:30, 45:15, 30:45 and 45:30 but more between these and the 0:60 ratio. Taking the F/TMP samples separately, function 2 also discriminates clearly between samples with addition ratios of 0:0 and 60:0, and the rest. This is explained by the fact that the F/TMP samples with added EVOO at highest concentrations scored higher for *creaminess* and lower for *adhesiveness*. Fat is a well-known enhancer of creaminess sensations [81], which covers oral tissue with a coating therefore reducing the friction between food and oral tissue. Starch break-down and the resulting loss in viscosity allow fat droplets to migrate from the bulk of the bolus to the surface where it provides additional lubrication [83].

Categorical Principal Components Analysis (CATPCA)

Table 10 shows the components extracted in each CATPCA analysis [FMP and F/TMP subgroups considered separately (n = 128)] for eleven sensory attributes with eigenvalues greater than unity (Kaiser's rule) [84]. Analogously, in the CATPCA analyses, one colour attribute (*off-colours*) was rejected for the model as its variance was zero.

For the FMP samples, CATPCA revealed that three principal components (PC-1 to PC-3) explained almost 70% of the variance between the data, although the first two components explained 56% of the total variance. In turn, for the F/TMP samples, four principal components (PC-1 to PC-4) explained almost 74% of the variance between the data, but the first two components explained only 44% of the total variance. Table 11 shows the rotated varimax PC loadings. The loading for each scored sensory attribute represents its relative contribution to a principal component. Sensory attributes with the highest loadings for each PC are marked in boldface, indicating that this PC is dominated by these particular attributes and, therefore the meaning of each component.

Figure 9 shows the projection of the scored sensory attributes on the rotated space defined by the first, second and third factorial axes for both the FMP and F/TMP samples. Total variation explained in Figure 9a is 69.2%, whereas total variation explained in Figure 9b is lower (59%). When the eleven sensory attributes were taken into account for the FMP samples, the first PC, which explains 28.55% of the total variance (Table 10), presents positive and quite high correlations, in descending order, with *creaminess* and *authentic colour* (Table 11). This first PC also presents negative and relatively high correlation with *fibrousness* (Figure 9a). This means that the greater either the *creaminess* or the *authentic colour*, the lower the *fibrousness* of the FMP samples. This is logical as *fibrousness* is a geometrical textural attribute reflecting the graininess perceived during preparation of the sample in the mouth prior to swallowing [20], whereas creaminess is a complex sensory attribute that is often considered an indicator of high, premium-quality and is critical for consumer acceptance of many food products [85].

Despite the fact that *creaminess* is a sensory characteristic highly coveted in a wide range of food products, it has not been clearly defined. Recently, it was defined as the combined perception of fat, smoothness and thickness by Bayarri et al. [45]. Creaminess mainly depends on the microstructure, on some surface properties, as well as on certain interactions between texture attributes and some flavours [86]. Certainly, FMP samples with added EVOO, either alone or blended with INL, scored highest for *creaminess* and presented softer mouthfeel, promoting positive effect on the *authentic colour* of the resulting products by increasing yellowness. PC-1 integrates *fibrousness*, one geometrical attribute, and *creaminess*, both associated with the arrangement of particles in the product, and surface roughness or smoothness properties of MP texture according to Szczesniak [87]. Hence, component 1 may be identified mainly with geometrical textural attributes which serve as an index of the *creaminess/fibrousness* of the FMP samples. Analogously, De Wijk et al. [81], using a trained panel, developed a set of sensory attributes to describe flavour, odour, mouthfeel and afterfeel of commercial vanilla custard desserts. Two main sensory dimensions were recognised in the resulting sensory space, one running from melting to thick and another one running from rough to creamy/fatty. Similar information was obtained from a group of consumers that described the perceptible differences on texture of eight samples of commercial vanilla dairy desserts of different brands by means of Free Choice Profile. In this case the first dimension was related with thick-fluid attributes and the second one with rough-smooth texture [88].

Table 10. Total variation explained by the categorical principal components (CATPCA) under the two CATPCA analyses carried out

	FMP samples		F/TMP samples	
Component	**Eigenvalue**	**Percent of variance**	**Eigenvalue**	**Percent of variance**
PC-1	3.14	28.55	2.63	23.95
PC-2	3.00	27.32	2.22	20.16
PC-3	1.47	13.37	1.64	14.87
PC-4	-	-	1.63	14.85

FMP: fresh mashed potatoes; F/TMP: mashed potatoes subjected to a freeze/thaw cycle.

Table 11. Rotated varimax principal components (PC) loadings of the sensory attributes

Sensory attribute	FMP samples			F/TMP samples			
	PC-1	PC-2	PC-3	PC-1	PC-2	PC-3	PC-4
Authentic colour (Aucol)	**0.788***	-0.223	0.268	**0.777**	0.184	0.343	0.208
Shine (Shi)	0.158	**-0.842**	0.151	0.407	-0.244	0.635	-0.172
Uniformity (Unicol)	0.662	0.157	0.366	**0.726**	-0.095	-0.240	0.039
Consistency (Con)	0.206	**0.823**	0.223	0.006	**0.938**	-0.080	0.063
Adhesivenes (Adh)	0.151	**0.749**	0.308	0.024	**0.909**	-0.069	0.054
Creaminess (Cre)	**0.876**	-0.110	-0.136	0.672	0.260	0.088	-0.327
Fibrousness (Fib)	**-0.747**	-0.049	0.374	-0.094	-0.544	-0.264	0.565
Sweetness (Swe)	0.108	-0.222	**-0.869**	-0.094	-0.104	-0.029	**-0.830**
Authentic taste (Autas)	0.079	-0.620	0.286	0.506	0.066	0.047	0.658
Off-taste (Offtas)	-0.684	-0.183	0.054	0.023	-0.042	**-0.875**	-0.054
Ingredient taste (Ingtas)	0.427	**-0.707**	0.314	**0.782**	-0.168	0.448	0.038

* The highest PC loadings for each factor are in boldface; FMP: fresh mashed potatoes; F/TMP: mashed potatoes subjected to a freeze/thaw cycle.

PC-2 (which explains over 27% of the total variance) was positively weighted by *consistency* and *adhesiveness* and negatively dominated by *shine* and *ingredient taste* (Table 11). This means that when the assessors detected more mechanical strength in the FMP products, they detected poorer *shine* and *EVOO taste*. In the FMP samples, component 2 was mainly a contrast between mechanical texture attributes (*consistency*, *adhesiveness*) and appearance and taste ones (*shine*, *ingredient taste*), acting as an index of the mechanical behaviour and the *shine* associated with particularly fat-related flavours. PC-3 was negatively weighted by *sweetness*. Therefore, this sensory taste attribute was unrelated to any other of the scored sensory attributes. On the other hand, intercorrelated sensory attributes with nearly equal loadings are assumed to represent some redundancy. *Consistency* and *adhesiveness* had nearly equal loadings.

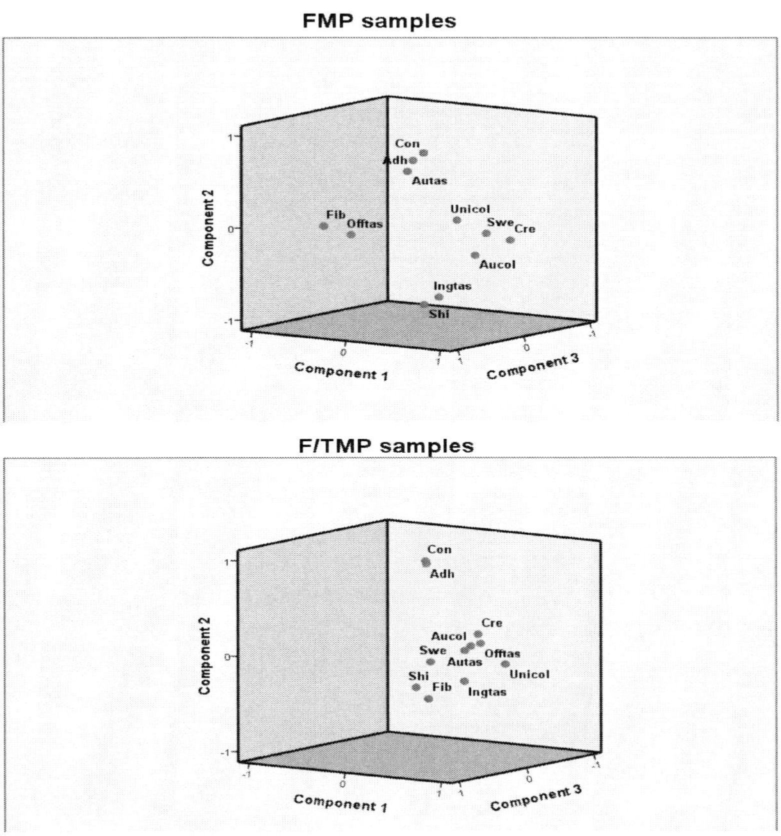

Figure 9. Plot for the first three principal components from categorical principal components analysis (CATPCA) carried out for: (a) FMP samples; (b) F/TMP samples; INL: inulin; EVOO: extra virgin olive oil.

In the case of the F/TMP samples (Figure 9b, Table 11), PC-1 (which explains over 24% of the total variance, Table 10) was positively weighted by *authentic colour*, *uniformity* and *ingredient taste*. The greater the addition of EVOO with respect to the frozen/thawed 0:0 control, the more intensive either the colour or the fat-related flavours. All the three loadings are positive and two variables are sensory appearance attributes. Therefore, component 1 may be interpreted as a representation of the presence of EVOO affecting either the colour or the taste of the F/TMP samples. It is worth noting that by considering the F/TMP samples separately, PC-2 is dominated by the same sensory texture attributes than in the case of their FMP counterparts (Table 11). Again, *consistency* and *adhesiveness* positive and highly weighted component 2, which may be identified chiefly with mechanical texture attributes. PC-3 and PC-4 presented a negative correlation with *off-taste* and *sweetness* respectively, indicating the absence of any relationship between these taste attributes and those well correlated under the first principal components plane (PC-1 and PC-2), and showing that any effect of *off-taste* or *sweet taste* detected in the product was not significant for the appearance and texture perceived in the same.

Reliability Analysis

Two or more items measuring the same thing for the same individual may not have exactly the same value. However, repeated measurements of a series of individuals will show some agreement or consistency [89]. In order to establish the levels of agreement between assessors, we used a reliability test. The reliability of the scores for each sensory property evaluated was tested by means of Cronbach's alpha coefficient [α] [29] (Table 12), based on the average inter-item correlations between its measures. The coefficient α is a measure of the internal consistency within the group of assessors (panel). Again, the colour attribute *off-colours* was discarded in the analysis as its variance was zero.

It is a widely accepted rule that α should be at least 0.70 in order to demonstrate consistence in the sub-items and know that the items are measuring the same construct [84]. The results obtained in the FMP samples showed a very good internal consistency for *ingredient taste*, *adhesiveness*, *consistency*, *off-taste* and *authentic colour* (Cronbach's α = 0.989, 0.979, 0.975, 0.959 and 0.945, in that order, respectively). Conversely, the assessors showed a very poor internal consistency for *sweetness* (Cronbach's α = -0.529). In turn, the results obtained in the F/TMP samples showed a very acceptable internal consistency for *off-taste*, *ingredient taste*, *consistency*, *adhesiveness*, *authentic colour* and *creaminess* (Cronbach's α = 1.000, 0.994, 0.979, 0.976, 0.960 and 0.948, in that order, respectively). However, again assessors showed a poorer internal consistency for *sweetness* (Cronbach's α = 0.657). Therefore, *ingredient taste*, *consistency* and *adhesiveness* were the three sensory attributes that showed a superior internal consistency in both FMP and F/TMP samples. The relatively low internal consistency detected for *sweetness* suggests that more training in the use of this taste attribute might be needed.

Table 12. Reliability analyses for FMP and F/TMP samples

Sensory attribute	Cronbach's alpha coefficient	
	FMP samples	F/TMP samples
Authentic colour	0.945	0.960
Shine	0.912	0.816
Uniformity	0.934	0.797
Consistency	0.975	0.979
Adhesiveness	0.979	0.976
Creaminess	0.931	0.948
Fibrousness	0.898	0.839
Sweetness	-0.529	0.657
Authentic taste	0.919	0.755
Off-taste	0.959	1.000
Ingredient taste	0.989	0.994

When a panel member is discarded a change is produced in the Cronbach alpha coefficient value. By studying this variation (data not shown), it is possible to determine the homogeneity of the judgments made by each panellists for each attribute. Thus, a decrease in the Cronbach coefficient indicates that the assessor is very homogeneous whereas an increase in the α value would indicate that the contrary is true. It was found that in the FMP samples, assessors 3 and 7 were the least homogeneous in their scores for the taste attributes *off-taste* and *ingredient taste*, whereas assessors 2, 4, 6 and 8 were the most homogeneous for all the sensory attributes. In turn, in the F/TMP samples, panellists 2 and 6 were the least homogeneous for all the sensory attributes, except for *sweetness*, while panellists 3 and 7 were the most homogeneous for all the sensory attributes.

CONCLUSION

The rheology of mashed potatoes is governed by a continuous phase (amylose/amylopectin matrix), which is eventually modified by the presence of INL/EVOO blends. In FMPA, FMPB and F/TMPB products, samples with no (INL+EVOO) blends exhibited greater elasticity than those with added INL/EVOO content. This indicates a shift in the viscoelastic behaviour of the samples with added INL:EVOO ratios towards more viscous-like products. INL crystallites and oil droplets should be regarded as inactive filler particles. The decrease in the G' and G'' values detected in the F/TMP samples as compared to their FMP counterparts is attributed to amylopectin retrogradation associated with the freeze/thaw cycle. However, the presence of hydrocolloids significantly improved freeze/thaw stability and mitigated the effect of INL:EVOO ratio variations on the viscoelasticity of the mashed potatoes. With respect to 0:0 controls, the addition of INL alone produced the lightest-coloured mashed potatoes (higher L^*/b^* ratio), whereas EVOO addition alone caused samples to darken (higher colour difference (ΔE^*)). When INL/EVOO blends were added, the darkening produced by EVOO addition was stronger than the lightening caused by INL incorporation. With regard to 0:0 controls, increasing yellowness due to the presence of EVOO improved product colour, as the flesh of Kennebec potatoes is too white.

In FMP samples considered separately, LDA showed that the most obvious discrimination was between 0:0 control and samples containing 0:60, 15:45 and 30:45 ratios, which is mainly ascribed to the fact that the samples containing higher EVOO concentrations received higher *ingredient taste* scores due to an aromatic note and fat-related flavours detected by the panellists. In F/TMP samples considered separately, LDA showed that the sensory characteristics of the product with an added 60:0 ratio were similar to those of the product without added ingredients, which is mainly attributed to the fact that the samples without added EVOO received higher and lower *adhesiveness* and *creaminess* scores respectively. Panellists also scored the F/TMP samples with an added 0:60 ratio higher than the rest for *ingredient taste*, indicating that a freeze/thaw cycle did not affect the aroma and taste attributes of the extra virgin olive oil.

CATPCA of the FMP samples, considered separately, showed that component 1 served as a *creaminess/fibrousness* index of the product in response to increased lubricating and coating properties produced by the presence of EVOO, whereas when the F/TMP samples were analysed separately, the first principal component served as an index of *authentic colour*, *uniformity* and *ingredient taste* in response to improved yellowness and flavour-releasing properties of the EVOO. The results obtained were highly consistent (Cronbach alpha coefficient) for all the sensory attributes except *sweetness*. The addition of either INL or EVOO at 30 g kg^{-1} is strongly recommended in order to achieve an increase in the daily intake of soluble fibre and monounsaturated fat (MUFA) while at the same time improving the sensory quality of the resulting products. The information obtained may be of great interest in designing mashed potatoes with nutritional and sensory characteristics that meet consumer demands.

ACKNOWLEDGEMENTS

The authors wish to thank the Spanish Ministry of Science and Innovation for financial support (AGL2007-62851). The authors also thank Laura Barrios (SGAI-CSIC) for assistance with non-parametric statistical tests.

REFERENCES

[1] Van Trijp, H. C. M., Punter, P. H., Mickartz, F., & Kruithof, L. (2007). The quest for the ideal product: comparing different methods and approaches. *Food Quality and Preference*, *18*, 729-740.

[2] Hasler, C. M. (1998). Functional foods: their role in disease prevention and health promotion. *Food Technology*, *52* (11), 63-70.

[3] Arcia, P. L., Costell, E., & Tárrega, A. (2010). Thickness suitability of prebiotic dairy desserts : relationship with rheological properties. *Food Research International*, *43*, 2409-2416.

[4] Arcia, P. L., Costell, E., & Tárrega, A. (2011). Inulin blend as prebiotic and fat replacer in dairy desserts : optimization by response surface methodology. *Journal of Dairy Science*, *94* (5), 2192-2200.

[5] Roberfroid, M. B., Van Loo, J., & Gibson, G. (1998). The bifidogenic nature of chicory inulin and its hydrolysis products. *Journal of Nutrition*, *128*, 11-19.

[6] Coussement, P. A. A. (1999). Inulin and oligofructose: safe intakes and legal status. *Journal of Nutrition*, *129* (7S), 1412S-1417S.

[7] Bendini, A., Cerretani, L., Carrasco-Pancorbo, A., Gómez-Caravaca, A. A., Segura-Carretero, A., Fernández-Gutiérrez, A., & Lercker, G.

(2007) Phenolic molecules in virgin olive oils: a survey of their sensory properties, health effects, antioxidant activity and analytical methods. An overview of the last decade. *Molecules*, *12*, 1679-1719.

[8] US Food and Drug Administration. Press Release P04-100. November 1, 2004. http://www.fda.gov/bbs/topics/news/2004/NEW01129.htlm. Accessed on April 14, 2010.

[9] Delgado, C., & Guinard, J.-X. (2011). Sensory properties of californian and imported extra virigin olive oils. *Journal of Food Science*, *76* (3), S170–S176.

[10] Boskou, D., Blekas, G., & Tsimidou, M. (2006). Olive oil composition. In D. Boskou (Ed.), *Olive oil: Chemistry and Technology*, *Vol. 4* (pp. 41-72), Champaign: AOCS Press.

[11] van Vliet, T., van Aken, A., de Jongh, H. H. J., & Hamer, R. J. (2009). Colloidal aspects of texture perception. *Advances in Colloid and Interface Science*, *150*, 27-40.

[12] Canet, W., & Alvarez, M. D. (2006). Quality and safety of frozen vegetables. In S. Da-Wen (Ed.), *Handbook of Frozen Food Processing and Packaging* (pp. 377-415), London/New York: Boca Raton: CRC Taylor & Francis Group.

[13] Wilkinson, C., Dijksterhuis, G. B., & Minekus, M. (2000). From food structure to texture. *Trends in Food Science and Technology*, *11*, 442-450.

[14] van Vliet, T. (2002). On the relation between texture perception and fundamental mechanical parameters for liquids and time dependent solids. *Food Quality and Preference*, *13*, 227-236.

[15] Adams, M. J, Bedford, L. V., & Geering, J. (1981). *QAV – a Method for the Sensory Appraisal of Quality Processed Vegetable Varieties. Technical Memorandum 278.* The Campden Food Preservation Research Association, Chipping, Campden, Gloucestershire, UK, pp 7-29.

[16] Canet, W., Fernández, C., & Alvarez, M. D. (2011). Functionality of amidated low methoxyl pectin and xanthan gum blends on the texture and color of mashed potatoes. *Journal of Food Process and Engineering*, *34*, 746-776.

[17] Fernández, C., Alvarez, M. D., & Canet, W. (2008). Steady shear and yield stress of fresh and frozen/thawed mashed potatoes: effect of biopolymers addition. *Food Hydrocolloids*, *22*, 1381-1395.

[18] Fernández, C., Canet, W., & Alvarez, M. D. (2010). Rheological properties and sensory attributes of mashed potatoes made with

mixtures of sodium caseinate and xanthan gum. In C. S. Hollingworth (Ed.), *Food Hydrocolloids: Characteristics, Properties and Structures* (pp. 81–117), New York: Nova Science Publishers, Inc.
[19] Alvarez, M. D., Fernández, C., & Canet, W. (2011). Flow behaviour and sensory properties of extra virgin olive oil-enriched mashed potatoes: influence of cryoprotectants and freezing. In N. Benkeblia, & Pedreschi, F. (Eds.), *Food 5 (Special Issue 1)* United Kingdom: Global Science Books (GBS).
[20] Alvarez, M. D., Canet, W., & Fernández, C. (2008). Effect of addition of cryoprotectants on the mechanical properties, colour and sensory attributes of fresh and frozen/thawed mashed potatoes. *European Food Research and Technology*, *226*, 1525-1544.
[21] Alvarez, M. D., Fernández, C., & Canet, W. (2009). Enhancement of freezing stability in mashed potatoes by the incorporation of kappa-carrageenan and xanthan gum blends. *Journal of the Science of Food and Agriculture*, *89*, 2115-2127.
[22] Fernández, C., Canet, W., & Alvarez, M. D. (2009). Quality of mashed potatoes: effect of adding blends of kappa-carrageenan and xanthan gum. *European Food Research and Technology*, *229*, 205-222.
[23] Alvarez, M. D., Canet, W., & Fernández, C. (2005). Effect of freezing/thawing conditions and long-term frozen storage on the quality of mashed potatoes. *Journal of the Science of Food and Agriculture*, *85*, 2327–2340.
[24] Navarro, A. S., Martino, M. N., & Zaritzky, N. E. (1997). Correlation between transient rotational viscometry and a dynamic oscillatory test for viscoelastic starch based systems. *Journal of Texture Studies*, *28*, 365-385.
[25] Mohamed, A., & Xu, J. (2003). Effect of ionic strength and pH on the thermal and rheological properties of soy protein-amylopectin blend. *Food Chemistry*, *83*, 227-236.
[26] Alvarez, M. D., Fernández, C., Solas, M. T., & Canet, W. (2011). Viscoelasticity and microstructure of inulin-enriched mashed potatoes: influence of freezing and cryoprotectants. *Journal of Food Engineering*, *102*, 66-76.
[27] O'Mahony, M. (1986). Additional nonparametric tests. In S. R. Tannenbaum, & P. Walstra (Eds.), *Statistical methods and procedures* (pp. 303–365). New York: Marcel Dekker Inc.
[28] ISO 8587. (2006). *Sensory Analysis. Methodology-Ranking*. International Organization for Standardization, Genève, Switzerland.

[29] Cronbach, L. (1951). Coefficient alpha and the internal structure of tests. *Psychometrica*, *16*, 297–334.

[30] Tunick, M. H. (2011). Small-strain dynamic rheology of food protein networks. *Journal of Agriculture and Food Chemistry*, *59*, 1481-1486.

[31] González-Tomás, L., Coll-Marqués, J., & Costell, E. (2008). Viscoelasticity of inulin-starch-based dairy systems. Influence of inulin average chain length. *Food Hydrocolloids*, *22*, 1372–1380.

[32] Campo-Deaño, L., Tovar, C. A., & Borderías, J. (2010). Effect of several cryoprotectants on the physicochemical and rheological properties of suwari gels from frozen squid surimi made by two methods. *Journal of Food Engineering*, *97*, 457-464.

[33] Kaur, L., Singh, J., McCarthy, O. J., & Singh, H. (2007). Physico-chemical, rheological and structural properties of fractionated potato starches. *Journal of Food Engineering*, *82*, 383–394.

[34] Clark, A. H., & Ross-Murphy, S. B. (1987). Structural and mechanical properties of biopolymer gels. Advances in Polymer Science, 83, 57–192.

[35] Rao, M. A., & Cooley, H. J. (1992). Rheology of tomato pastes in steady and dynamic shear. *Journal of Texture Studies*, *23*, 415–425.

[36] Ahmed, J., & Ramaswamy, H.S. (2006). Viscoelastic properties of sweet potato puree infant food. *Journal of Food Engineering*, *74*, 376–382.

[37] Arocas, A., Sanz, T., & Fiszman, S. M. (2009). Improving effect of xanthan and locust bean gums on the freeze-thaw stability of white sauces made with different native starches. *Food Hydrocolloids*, *23*, 2478-2484.

[38] Ferrero, C., Martino, M. N., & Zaritzky, N. E. (1994). Corn starch–xanthan gum interaction and its effect on the stability during storage of frozen gelatinized suspensions. *Starch*, *46*(8), 300–308.

[39] Rodríguez-Hernández, A. I., & Tecante, A. (1999). Dynamic viscoelastic behavior of gellan-i-carrageenan and gellan-xanthan gels. *Food Hydrocolloids*, *13*, 59–64.

[40] Sun, C., & Gunasekaran, S. (2009). Effects of protein concentration and oil-phase volume fraction on the stability and rheology of menhaden oil-in-water emulsions stabilized by whey protein isolate with xanthan gum. *Food Hydrocolloids*, *23*, 165-174.

[41] Dickinson, E., & Chen, J. S. (1999). Heat-set whey protein emulsion gels: Role of active and inactive filler particles. *Journal of Dispersion, Science and Technology*, *20*, 197–213.

[42] Alvarez, M. D., Fernández, C., Olivares, M. D., & Canet, W. (2010). Rheological behaviour and functionality of inulin-extra virgen olive oil-based mashed potatoes. *International Journal of Food Science and Technology*, *45*, 2108-2118.

[43] Jampen, S., Britt, I. J., Yada, S., & Tung, M. A. (2001). Rheological properties of gellan gels containing filler particles. *Journal of Food Science*, *66*, 289–293.

[44] Zimeri, J. E., & Kokini, J. L. (2003). Morphological characterization of the phase behaviour of inulin-waxy maize starch systems in high moisture environments. *Carbohydrate Polymers*, *52*, 225-236.

[45] Bayarri, S., González-Tomás, L., Hernando, I., Lluch, M. A., & Costell, E. (2011). Texture perceived on inulin-enriched low-fat semisolid dairy desserts. Rheological and structural basis. *Journal of Texture Studies*, *42*, 174-184.

[46] van Vliet, T. (1988). Rheological properties of filled gels: Influence of filler matrix interaction. *Colloid and Polymer Science*, *266*, 518-524.

[47] Sosa-Herrera, M. G., Berli, C. L. A., & Martínez-Padilla, L. P. (2008). Physocochemical and rheological properties of oil-in-water emulsions prepared with sodium caseinate/gelan gum mixtures. *Food Hydrocolloids*, *22*, 934-942.

[48] Eliasson, A. C., & Kim, H. R. (1992). Changes in rheological properties of hydroxypropyl potato starch pastes during freeze-thaw treatments I. A rheological approach for evaluation of freeze-thaw stability. *Journal of Texture Studies*, *23*, 279-295.

[49] Guggisberg, D., Cuthbert-Steven, J., Piccinali, P., Bütikofer, U., & Eberhard, P. (2009). Rheological, microstructural and sensory characterization of low-fat and whole milk set yogurt as influenced by inulin addition. *International Dairy Journal*, *19*, 107-115.

[50] Hébette, C. L. M., Delcour, J. A., Koch, M. H. J., Booten, K., Kleppinger, R., Mischenkod, N., & Reynaers, H. (1998). Complex melting of semi-crystalline chicory (*Cichoriumintybus L.*) root inulin. *Carbohydrate Research*, *310*, 65–75.

[51] Franck, A., & De Leeheer, L. (2005). Inulin In A. Steinbüchel, & S. Ki Ehee (Eds.), *Polysaccharides and Polyamides in the Food Industry*, (pp. 281-322), Weinheim: Wiley-VCH Verlag GmbH & Co. KGaA.

[52] Giannouli, P., & Morris, E. R. (2003). Cryogelation of xanthan. *Food Hydrocolloids*, 17, 495-501.

[53] Chantrapornchai, W., Clydesdale, F., & McClements, D. J. (1999). Influence of droplet characteristics on the optical properties of colored oil-in-water emulsions. *Colloids and Surfaces A*, *155*, 373-382.
[54] Criado, M. N., Romero, M. P., Casanovas, M., & Motilva, M. J. (2008). Pigment profile and colour of monovarietal virgin olive oils from Arbequina cultivar obtained during two consecutive crop seasons. *Food Chemistry*, *110*, 873-880.
[55] Bot, A., Erle, U., Vreeker, R., & Agterof, W. (2004). Influence of crystallisation conditions on the large deformation rheology of inulin gels. *Food Hydrocolloids*, *18*, 547-556.
[56] O'Leary, E., Gormley, T. R., Butler, F., & Shilton, N. (2000). The effect of freeze-chilling on the quality of ready-meal components. *Lebensmittel-Wissenschaft und-Technologie*, *33*, 217–224.
[57] González-Tomás, L., Bayarri, S., & Costell, E. (2009). Inulin-enriched dairy desserts: physicochemical and sensory aspects. *Journal of Dairy Science*, *92*, 4188-4199.
[58] Akalin, A. S., Karagözlü, C., & Ünal, G. (2008). Rheological properties of reduced-fat and low-fat ice cream containing whey protein isolate and inulin. *European Food Research and Technology*, *227* (3), 889-895.
[59] Castro, F. P., Cunha, T. M., Barreto, P. L. M., Amboni, R. D. M. C., & Prudêncio, E. S. (2009). Effect of oligofructose incorporation on the properties of fermented probiotic lactic beverages. *International Journal of Dairy Technology*, *62* (1), 68-74.
[60] Kokini, J. L., & Cussler, E. L. (1983). Predicting the texture of liquid and melting semi-solid foods. *Journal of Food Science*, *48* (4), 1221-1225.
[61] Cussler, E. L., Kokini, J. L., Weinheimer, R. L., & Moskowitz, H. R. (1979). Food texture in mouth. *Food Technology*, *33* (10), 89-92.
[62] Frost, M. B., Dijksterhuis, G., & Martens, M. (2001). Sensory perception of fat in milk. *Food Quality and Preference*, *12* (5-7), 327-336.
[63] Kilkast, D., & Clegg, S. (2002). Sensory perception of creaminess and its relationship with food structure. *Food Quality and Preference*, *13* (7, 8), 609-623.
[64] de Wijk, R. A., Terpstra, M. E. J., Janssen, A. M., Prinz, J. F. (2006). Perceived creaminess of semi-solid foods. *Trends in Food Science & Technology*, *17*, 412-422.

[65] de Wijk, R. A., Prinz, J. F., & Janssen, A. M. (2006). Explaining perceived oral texture of starch-based custard desserts. *Food Hydrocolloids*, *20*, 24-34.
[66] Jellema, R. H., Janssen, A. M., Terpstra, M. E. J., De Wijk, R. A., & Smilde, A. K. (2005). *Journal of Chemometrics*, *19*, 191-200.
[67] Villegas, B., Carbonell, I., & Costell, E. (2007). Inulin milk beverages: sensory differences in thickness and creaminess using r-index analysis of the ranking data. *Journal of Sensory Studies*, *22*, 377-393.
[68] Kip, P., Meyer, D., & Jellema, R. H. (2006). Inulins improve sensoric and textural properties of low-fat yoghurts. *International Dairy Journal*, *16*, 1098-1103.
[69] Modzelewska-Kapituła, M., & Kłębukowska, L. (2009). Investigation of the potential for using inulin HPX as a fat replacer in yoghurt production. *International Journal of Dairy Technology*, *62*, 209-214.
[70] Lobato, L. P., Grossmann, M. V. E., & Benassi, M. T. (2009). Inulin addition in starch-based dairy desserts: Instrumental texture and sensory aspects. *Food Science and Technology International*, *15*, 317-323.
[71] Tárrega, A., & Costell, E. (2006). Effect of inulin addition on rheological and sensory properties of fat-free starch-based dairy desserts. *International Dairy Journal*, *16*, 1104-1112.
[72] Franck, A. (2002). Technological functionality of inulin and oligofructose. *British Journal of Nutrition*, *2*, 287–291.
[73] Angerosa, F., Servili, M., Selvaggini, R., Taticchi, A., Esposto, S., & Montedoro, G. (2004). Volatile compounds in virgen olive oil: occurrence and their relationship with the quality. *Journal of Chromatography A*, *1054* (1), 17-31.
[74] Cerretani, L., Salvador, M. D., Bendini, A., & Fregapane, G. (2008). Relationship between sensory evaluation performed by Italian and Spanish official panels and volatile and phenolic profiles of virgin olive oils. *Chemosensory Perception*, *1*, 258-267.
[75] Cardarelli, H. R., Aragon-Alegro, L. C., Alegro, J. H. A., Castro, I. A. & Saad, M. I. (2008). Effect of inulin and lactobacillus paracasei on sensory and instrumental texture properties of functional chocolate mouse. *Journal of the Science of Food and Agriculture*, *88*, 1318-1324.
[76] Alvarez, M. D., Fernández, C., Jiménez, M. J., & Canet, W. (2011). Texture of extra virgin olive oil-enriched mashed potatoes: sensory, instrumental and structural relationships. *Journal of Texture Studies*, *42*, 413-429.

[77] Dresselhuis, D. M., Cohen Stuart, M. A., van Aken, G. A., Schipper, R. G., & de Hoog, E. H. A. (2008). Fat retention at the tongue and the role of saliva: adhesion and spreading of 'protein-poor' versus 'protein-rich' emulsions. *Journal of Colloid and Interface Science*, *321*, 21-29.

[78] Dresselhuis, D. M., de Hoog, E. H. A., Cohen Stuart, M. A., Vingerhoeds, M. H., van Aken, G. A. (2008). The occurrence of in-mouth coalescence of emulsion droplets in relation to perception of fat. *Food Hydrocolloids*, *22*, 1170-1183.

[79] Benjamins J., Vingerhoeds, M. H., Zoet, F. D., de Hoog, E. H. A., van Aken, G. A. (2009). Partial coalescence as a tool to control sensory perception of emulsions. *Food Hydrocolloids*, *23*, 101-115.

[80] Taylor, A. J. (2002). Release and transport of flavors in vivo: physicochemical, physiological, and perceptual considerations. *Comprehensive Reviews in Food Science and Food Safety*, *1*, 45-57.

[81] de Wijk, R. A., van Gemert, L. J., Terpstra, M. E. J., & Wilkison, C. L. (2003). Texture of semi-solids; sensory and instrumental measurements on vanilla custard desserts. *Food Quality and Preference*, *14*, 305-317.

[82] Kim, Y., Faqih, M. N., & Wang, S. S. (2001) Factors affecting gel formation of inulin. *Carbohydrate Polymers*, *46*, 135-145.

[83] de Wijk, R. A., & Prinz, J. F. (2007). Fatty versus creamy sensations for custard desserts, white sauces, and mayonnaises. *Food Quality and Preference*, *18*, 641-650.

[84] Ojiako, U., Ashleigh, M., Chipulu, M., & Maguire, S. (2011). Learning and teaching challenges in project management. *International Journal of Project Management*, *29*, 268-278.

[85] Antmann, G., Ares, G., Salvador, A., Varela, P., Fiszman, S. M. (2011). Exploring and explaining creaminess perception: consumers' underlying concepts. *Journal of Sensory Studies*, *26*, 40-47.

[86] Weene, H., Jellema, R. H., & De Wijk, R. A. (2005). Sensory sub-attributes of creamy mouthfeel in commercial mayonnaises, custard desserts and sauces. *Food Quality and Preference*, *16*, 163-170.

[87] Szczesniak, A. S. (2002). Texture is a sensory property. *Food Quality and Preference*, *13*, 215-225.

[88] González-Tomás, L., & Costell, E. (2006). Relation between consumer's perception of color and texture of dairy desserts and instrumental measurements using GPA. *Journal of Dairy Science*, *89,* 4511-4519.

[89] Guerrero, L., Colomer, Y., Guàrdia, M. D., Xicola, J., & Clotet, R. (2000). Consumer attitude towards store Brands. *Food Quality and Preference*, *11*, 387-395.

In: Inulin
Editor: Michael G. Marlow
ISBN: 978-1-63463-985-9

Chapter 2

PROCESSING PLANTS CONTAINING INULIN

Václav Veselý[1]*, Jana Budovičová[2], Jiří Hanika[1], Miroslav Punčochář[1] and Martin Bárnet[3]

[1]Institute of Process Fundamentals, Czech Academy of Sciences, Prague, Czech Republic
[2]Institute of Chemical Technology, Prague, Czech Republic
[3]AGRA GROUP a.s., Střelské Hoštice, Czech Republic

1. PROCESSING OF STEMS AND TUBERS OF JERUSALEM ARTICHOKE

Abstract

Processing of tubers of the Jerusalem artichoke (further on JA) to syrup containing inulin, fructose, ethyl alcohol and lactic acid is discussed. Dry matter is produced by drying of tubers so they can be ground into powder and used as a food additive. It does not contain gluten and is thus suitable for people with celiac disease. Fructose syrup is produced by acidic hydrolysis of tuber juice. The method of hydrolysis, which is based on electrolytic transference of ions and the processing of fructose syrup with its subsequent fermentation to ethyl alcohol and lactic acid have been tested. Hydrolysis of protein obtained by pasteurisation of tuber juice of JA has been verified.

* vesely@icpf.cas.cz.

Keywords: Jerusalem artichoke, yacón, hydrolysis, fuel ethanol, lactic acid, inulin, biotechnology

1.1. Introduction

Inulin is a storage polysaccharide, generally formed by a linear chain of several molecules of D-fructofuranose and one molecule of D-glucose, although this arrangement is not always valid. Chemically, inulin is a type of fructan, that is, a linear polydisperse carbohydrate. Fructofurane molecules differ depending on the type of plant and its ripeness. Usually the number of molecules of fructose is 5 to 60. Inulin is not a chemical individual and its composition is also specified by its degree of polycondensation (DP), which corresponds to the ratio of molecules of fructose and glucose. Inulin is soluble in water and its solubility is dependent on temperature.

The raw materials most commonly used for industrial production of inulin are chicory (*Cichorium intybus*) and Jerusalem artichoke (*Helianthus tuberosus*) (JA). Inulin extracted from chicory is commercially available as Raftiline, produced by the Belgian company Orafti. Other raw materials for production of inulin include yacón, dahlia, garlic and onion. Inulin, used for the storage of energy, is located in the underground part of plant, and is extracted by pressing fresh juice from the tubers. The advantage of JA is the fact that tubers are able to survive temperatures as low as -20°C, which means that they can be cultivated in a wide range of earth zones. Plants can grow to between two and three metres in height; their vegetation time is between four and six months and they blossom with small yellow flowers from August to October. They are grown in a similar way to potatoes, and harvested in the autumn after drying of the stems, or in the spring after the soil defrosts. If the weather is favourable, JA tubers can be harvested throughout this entire period. Usual yield of JA tubers is 30-40 tons/ha. The yield of the above-ground part is 10-15 tons/ha. JA is cultivated as a perennial plant and propagated vegetatively. Tubers are often wrinkled, which becomes disadvantageous at the cleaning stage.

Chicory is cultivated as an annual plant, propagated by seeds and thus planted each year. The main product is the root, which is conical in shape, white, and relatively smooth. The yield is similar to that of JA, i.e., approximately 20-40 tons/ha. The above-ground part is usually not utilised. Chicory is harvested in the autumn, and the root is not frost-resistant.

Another plant which can be industrially utilised for the production of inulin is yacón. This also forms smooth underground tubers, and is harvested in the autumn with yields of 15-25 tons/ha. The underground part can be utilised in a similar way to JA, but its tubers are not frost-resistant. They are propagated vegetatively, with the pre-cultivated plants planted in spring. Yacón tubers are large and smooth, and can reach a weight of up to 2kg. Yacón and JA tubers and chicory roots are presented in Figure 1.

Yacón tubers

Jerusalem artichoke tubers

Chicory roots

Figure 1. Tubers of yacón, and Jerusalem artichoke and roots of chicory.

For the industrial production of inulin from JA tubers it is necessary to consider not only the content of active compound in this raw material, but also the method by which the tubers are cleaned and washed. Proper cleaning is one of the most important steps in the processing of tubers. Thus, only those JA tubers with shapes and dimensions convenient for washing can be used in the process. This could be a disadvantage. On the other hand, one significant advantage of JA is that their above-ground mass, i.e., the stems and leaves, can be used in agriculture.

Inulin content of JA and yacón tubers and chicory roots is in the range of 12-18 mass%. The difference among them is not significant. Chicory is currently the most frequently utilised raw material for the production of inulin. It is easily washed and contains large quantities of inulin, and its harvesting is well developed.

Inulin is broken down by the enzyme inulinase. These enzymes are not present in the human digestive tract, which is why this type of polysaccharide cannot be digested in a similar way to (e.g.) starch. However, inulin is broken down by *Bifidus lactis* bacteria, which, in conjunction with the peristalsis of the digestive tract (large intestine), results in probiotic effects.

Utilisation of tubers depends first of all on market demands. Various products are given in the literature [1-3]. For possible uses of the stems and tubers of JA, see Figure 3.

Stems, including leaves, can be utilised as fodder for cattle, but they must first be crushed and then fermented. Stems have numerous sharp and thorny outgrowths on their surface which prevent their direct utilisation as fodder. But they can be used as biomass for fermentation to biogas, which in turn can be utilised as a gaseous fuel for motors and turbines. Most commonly, however, the stems are dried after the vegetation period and used as solid fuel.

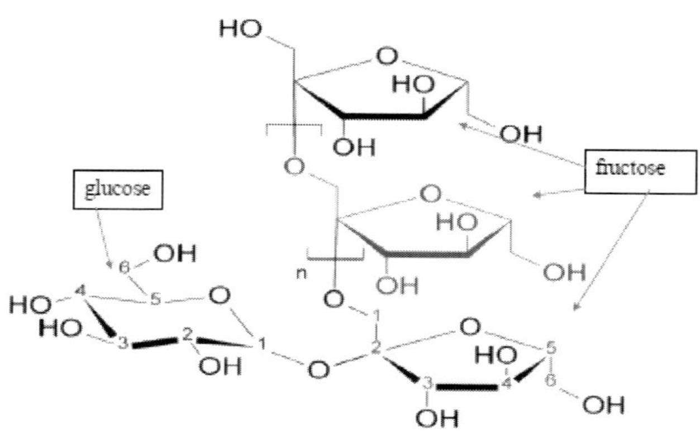

Figure 2. One structural formula of inulin.

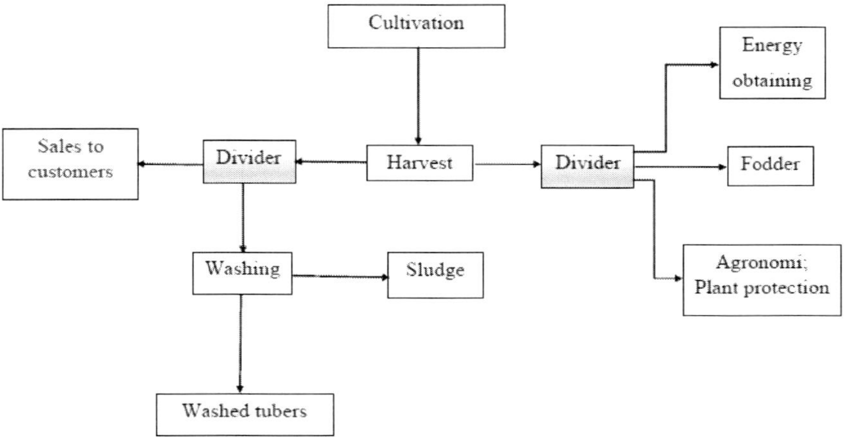

Figure 3. Diagram of the cultivation and utililsation of JA.

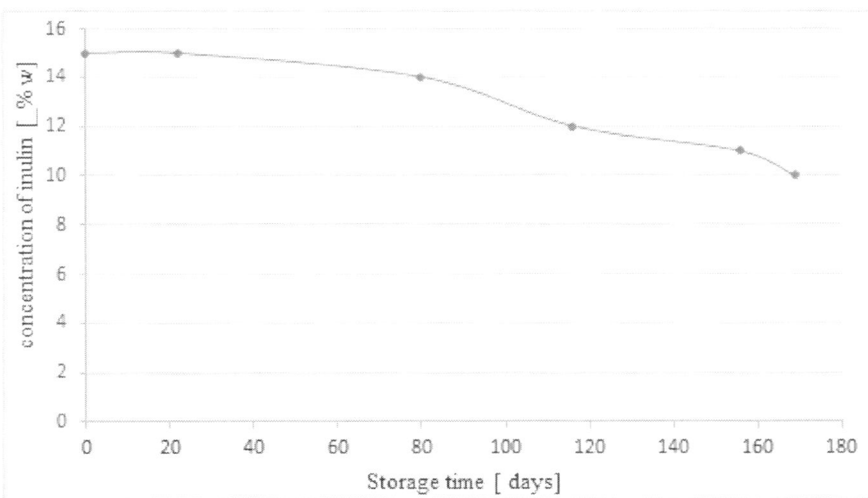

Figure 4. Storage of JA tubers; T=5°C.

After harvest, tubers are either stored or immediately processed. Tubers dry easily and their quality for direct consumption deteriorates. For these reasons, tubers must be stored at relatively high relative humidity and low temperature. Tubers could be well preserved for up to three months in dark and enclosed space at a temperature of around 5°C, but only until April, as then they start to germinate. Inulin in tubers begins to diminish by enzymatic breakage. Reduction of inulin content in tubers was recorded starting in November (time of harvest) up to May (time when tubers start to germinate). Tubers were kept in evacuated PE bags in a refrigerator at 5°C without light. For analysis, tubers were washed and then juice pressed out and analysed. Results are given in Figure 4.

All tubers must be properly washed before processing. Machine washing causes some loss of partially soluble polysaccharides and damages tuber surface. Damaged places become the target of contaminating organisms which results in rapid deterioration of the durability of the tubers. This is the reason why tubers must be processed immediately after washing. They could be used immediately as feed, or for production of polysaccharides, by pressing or leaching. Another possibility is the drying of the pulp of tubers with their subsequent processing.

1.2. Combustion, Torrefaction and Fermentation of Stems

During the vegetation period, yacón and Jerusalem artichoke form a significant above-ground part which can be utilised. After termination of the vegetation period this part of the plant dries. The leaves do not fall off, but they are only a small part of the dry matter, the main part of which is composed of the stems. These could either be incinerated or treated by torrefaction (low temperature pyrolysis). Elementary analysis of samples of dry, crushed stems is given in Table 1.

Table 1. Analysis of samples

Elements	mass%	mol%
C	41.7	3.475
H	5.81	5.81
N	0.98	0.07
S	0	
O	44.51	2.781875
Water (moisture)	0	
Ash	7	

Stems do not contain any sulphur and contain a very low concentration of nitrogen. Content of nitrogen is first of all given by content of vegetable proteins present in the stems. Analysis of carbon, hydrogen and oxygen corresponds to composition of hemicellulose $[C_6H_{10}O_5]_n$, which is very reactive, decomposes at a low temperature and burns easily. This corresponds with results of the GT and DTA analyses. The results are given in Figure 5.

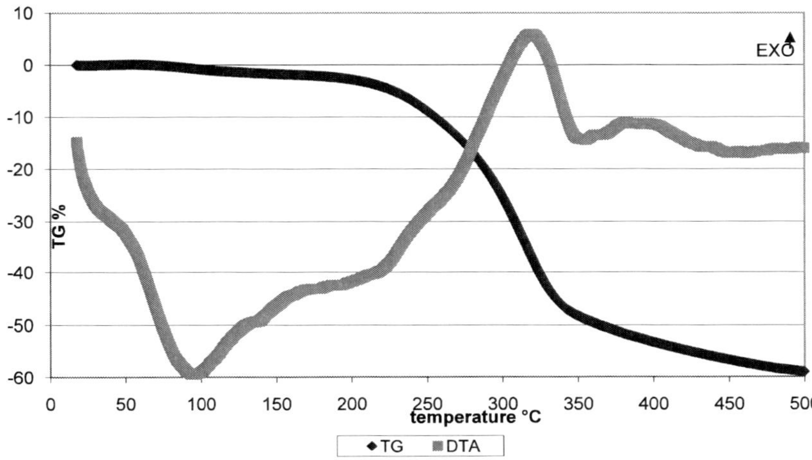

Figure 5. TG and DTA analysis of stems.

Two peaks are obvious on the DTA curve. At 100°C total evaporation of water takes place. This concerns about 1 mass%. At 200°C the exo-effect grows, reaching a maximum at 320°C. This is the decomposition of hemicellulose which is the most reactive component. It decomposes between 225 and 325°C. Another small exo-effect appears at 380-400°C, which corresponds to decomposition of cellulose. The loss of mass takes place for 40 relative % after decomposition of hemicellulose.

1.2.1. Combustion of Stems

Crushed stems were modified to briquettes by pressing. They had a diameter of about 66mm and were between 50-70 mm in height. Specific density of the briquettes was between 830 and 840 kg/m^3. Such briquettes were incinerated in a laboratory furnace in the stream of air at temperatures of between 950 and 1000°C. Content of CO, CO_2, oxygen, NO_x and sulphur dioxide in flue gases was measured by standard procedures after drying. Concentration was measured by a standard, calibrated NDIR analyser. Concentration of pollutants in flue gases in the incineration experiment was obtained and is shown in Table 2.

Table 2. Concentration of pollutants in flue gas

O_2	8.1	% vol.
CO	170	Ppm
NO_x	390	Ppm

Content of SO_2 was lower than 10 ppm. Ash after combustion contained 7.3 mass% of calcium, 5.26 mass% of potassium and 0.85 mass% of phosphorus. After combustion of the briquette, ash formed a grey deposit and did not split into pieces. See Figure 6.

Figure 6. Ash obtained by combustion of a briquette of Jerusalem artichoke stems.

No sulphuric compounds formed at combustion, but nitrogen oxides did. It is probable that oxides formed during destruction of the protein component in biomass.

1.2.2. Torrefaction

Recommended conditions for torrefaction (drying with removal of easily volatile compounds) were specified on the basis of GT and DTA curves for the specimen of crushed stems of yacón and Jerusalem artichoke. A common feature of crushed stems is their similar chemical composition, low ash content and zero content of organic chlorinated substances. [3] They contain various proportions of cellulose, hemicellulose, lignin, proteins, polysaccharides, starch and various organic acids, as well as biogenic inorganic elements, namely potassium, phosphorus, iron, magnesium and calcium. In addition, small amounts of sulphur and silicon were also detected. Calorific value of such waste materials ranges from 6 to 10 MJ/kg, depending on their moisture content. Hemicellulose is the most reactive component, which decomposes at temperatures of between 225-325°C, cellulose at 305-375°C and lignin at 250-500°C. [4] The proteins which are the main source of organically bound nitrogen are other, less stable substances.

Torrefaction leads to energetic utilisation of biomass due to elimination of moisture and partial low-temperature pyrolysis at temperatures ranging from 220-260°C. No tarry substances form under these conditions, but a substantial weight loss is reported. It was experimentally proven that weight drops by between 40 and 60wt%. Not only does this effect have a significant impact on logistics costs, but the material can be subsequently pelletised. This form of thermal pre-treatment unifies size and composition; the heating value of residuals is boosted, and material from various plants can be mixed and pelletised. [5] The pellets are dark, and given the content of polysaccharides and lipids the pellets are highly coherent and resistant to disintegration during storage or manipulation. The heating value of torrefied fuel increased from 16 to 20 MJ/kg. The substance formed is environmentally friendly fuel, the combustion of which does not produce NO_x or persistent organic pollutants.

The effect of the size of crushed stem particles on the torrefaction process was monitored. The charge of crushed stems was heated in a laboratory 2.5 l reactor by external heating. The reactor was supplied with nitrogen as a protective atmosphere. The temperature of the charge was measured by a thermocouple and in all experiments ranged from 235°C to 250°C. The temperature was kept constant for 20 minutes. After heating, the charge was cooled with nitrogen and the content of C, H and N in the torrefied chips was determined. The size of chips was monitored by sieving them through calibrated mesh. For the results see Table 3 below.

Table 3. Impact of chip size on elementary analysis in torrefied stems

Dp	C	H	N
[mm]	[% wt]	[% wt]	[% wt]
0-2	44.2	5.6	0
3-5	44	5.6	0.1
10-12	41.8	5.6	0.81

As expected, small chips were decomposed best, but as a crust formed on their dosing, the experiments were continued with chips that fell through a 5mm mesh.

The stem crushing process was optimised based on the previously obtained results. The effect of torrefaction of pellets made of raw material and wood chips was tested. The torrefied product was again used for production of pellets.

Figure 7. Pellets of torrefied stems.

It was conclusively proven that it was sensible to pelletise torrefied chips, otherwise the pellets of pure dried stems lost their coherence and crumbled.

The effect of the heating time needed was tested, at the temperature of 250°C, on the content of carbon and hydrogen in dry mass. The fuel heating value and its dependence on residence time in the reactor was calculated. The minimum estimated heating value could be calculated using Mendeleev´s formula based on elementary analysis as follows:

$$H_u = 33.913.m_C + 102.995.m_H - 10.885.(m_O - m_S) - 2.51m_{H2O} \quad (1)$$

Subsequently, the fuel heating value and its dependence on the time for which it had been kept at the given temperature of 250°C in the reactor (see Figure 7) was calculated.

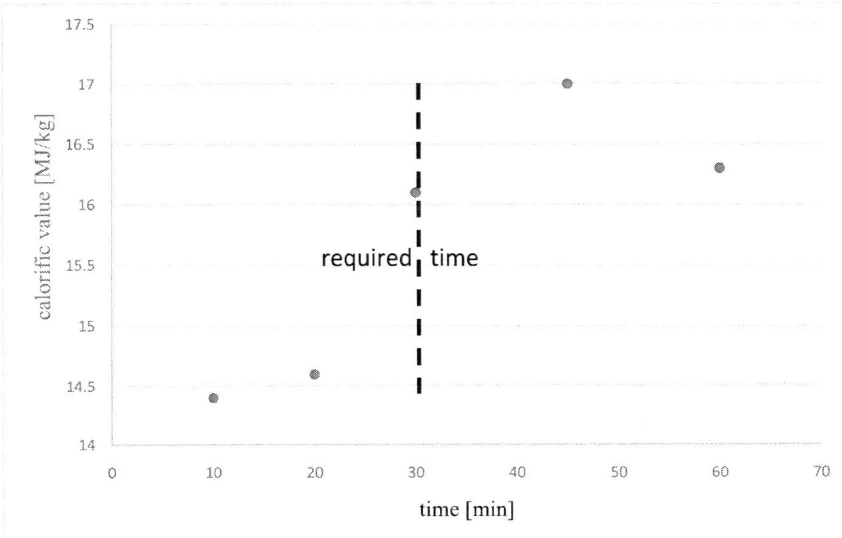

Figure 8. Calorific value in dependence on time of torrefaction. T=250°C.

The economic benefit of torrefaction is that the power necessary for leaching artichoke cuttings in water is obtained by combusting torrefied crushed stems. Should it be feasible to torrefy the plants in the field (where they have been harvested), additional logistics cost savings could be expected since the weight of stored and torrefied chips would be only half of the original weight, thus significantly reducing the transport cost. In addition, fuel made by torrefaction may be pelletised, and does not produce NO_X when combusted.

1.3. Production of Biogas by Fermentation

Because stems contain hemicelluloses and other polysaccharides it is possible to utilise this material for fermentation using methanogenic bacteria for production of fuel in the biogas station. The biogas produced contains methane (40-45%), carbon dioxide (35-45%) and water (10-15%). Low quantities of ammonia are present, with traces of hydrogen sulphide and siloxanes. The largest quantity of methane forms at fermentation from the decomposition of polysaccharides, lipids and proteins. Stems containing polysaccharides are a suitable additive of biomass for fermentation and thus for the production of biogas. Larger hemicellulose content prolongs acidic hydrolysis and the following phase of decomposition to biomass, which leads to the prolongation of residence time of biomass in the fermenter. The process of anaerobic fermentation of material could be blocked by (e.g.) trace quantities of undesirable compounds that suppress the origination and growth of fermentation bacteria, or by improper previous processing. This may concern residuals after alkaline or acidic hydrolysis etc. The composition of biomass depends not only on the composition of the original substrate but also on the production process used. Biomass can be also modified, e.g., by the Pressure Swing Adsorption (PSA) process or membrane technology, by which an energetically rich gas could be obtained. After purification, such gas can be utilised as fuel in motors and gas turbines. At combustion of biomass in a cogeneration unit which produces heat and electricity, about 65% of the energy of the biomass could be utilised. If waste heat is not efficiently utilised, this yield is reduced to just 30-35%.

1.4. Activities of Extracts in Agronomy

Bioactive substances can be obtained through the extraction of Jerusalem artichoke leaves and used in the production of botanic pesticides. Our experiments showed that substances obtained from leaves using methanol or butanol provide very good insecticidal effects.

For example, the tests demonstrated that extracts obtained from JA leaves of the white peel variety using methanol, butanol, acetone and ethyl acetate affect mortality and third instar larval growth inhibition in *Spodoptera littoralis* larvae. Dry plant mass was homogenised by grinding and macerated in organic solvents of various polarities for 48 hours at ambient temperature. The extracts obtained were filtered and thickened in a vacuum, and mixed into the diet (Stonefly Industries, Bryan, TX, USA) in the maximum dosage of 25 $mg.g^{1}$ of the diet. Observed characteristics included chronic toxicity, i.e., mortality achieved on day 5 from the start of the experiment, and weight gains. Lethal doses were estimated in further biological experiments for extracts showing mortality higher than 50% or growth inhibition.

Growth inhibition was associated with chronic mortality. Selected results (lethal doses) are shown in Table 4.

The extracts caused significant growth inhibition of *S.littoralis* larvae. Extracts obtained using n-butanol and methanol provided the most significant growth inhibition where larval growth was slower by 71.5% and 62.8% respectively, compared to the control group.

Table 4. Insecticidal activity of extracts from Jerusalem artichoke

	MeOH	25 mg/g – n-BuOH	25 mg/g – AcOH	25 mg/g – EtAc
Growth inhibition (%) at 25 $mg.g^{-1}$ **	62.8±5.6	71.5±7.3	17.6±5.2	26.5±8.3
Mortality (%) at 25 $mg.g^{-1}$**	80.6±3.8	98.5±5.7	15.7±2.9	49.6±3.2
LD_{50} (CI_{95})* - Mortality	18.8 (15.3-21.5)	3.8 (2.8-4.2)	>25	>25
LD_{50} (CI_{95})* – Growth inhibition	15.3 13.2-16.9)	1.2 (0.7-1.7)	>25	>25

* LD_{50} value in $mg.g^{-1}$ and CI_{95}–95% confidence intervals, compounds activity is considered significantly different when the 95% CI fail to overlap. **Average mortality or growth inhibition in % obtained after application of maximal dose 25 $mg.g^{-1}$.

2. Processing of Washed Tubers

Abstract

Cleaned tubers were cut and then fermented to lactic acid. During fermentation inulin remained in the digestate and did not decompose. Tuber chips can be dried and used as a natural dietary fibre for human beings or in mixture as feed for animals. Dry matter contained a variable amount of inulin. Raw juice was obtained by pressing and had to be pasteurised otherwise it started to ferment rapidly.

2.1. Introduction

Processing of washed tubers starts with their cutting into thin slices (cuts) or pressing in order to obtain raw juice. To avoid rapid decomposition of the raw juice by enzymes the juice must be sterilised at 80-100°C. Under these conditions, proteins in the juice coagulate. Once filtered, the sterile juice may be stored as a beverage or further processed. Proteins are separated by filtration, and then hydrolysed to amino acids in the presence of hydrochloric acid. Tuber slices were dried and the drying rate measured. The drying temperature influences oxidation of the slices and drying conditions must be adjusted according to the purpose of use. If the dried slices are to be used as flour, they must be dried at low temperature or antioxidants must be used. If the drying air is warmer than 300°C, the dried material is caramelised and may be used as one of components of coffee substitutes.

Figure 9 shows the main activities that may be pursued when processing washed tubers containing inulin. This is a general scheme and may be adjusted according to the specific processing conditions. Some processing procedures have been verified for tubers of JA and yacón. Processing of chicory roots is industrially well known, and therefore not the subject of this study. But we are certain that the general processing procedures are applicable for chicory as well.

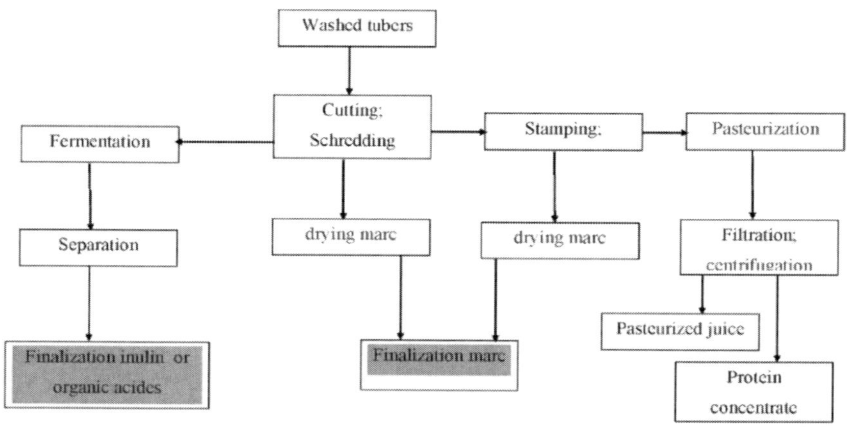

Figure 9. Schematic processing of washed JA tubers.

2.2. Drying of Tubers

Inulin as reserve polysaccharide is located in the underground part of the plant (tubers or roots). It is difficult to store the fresh tubers due to the fact that, besides inulin, the plant contains large quantities of water (70-80% v.) and enzymes which not only form polysaccharides but also enzymes that decompose the polysaccharides. Storage and utilisation of tubers in a dry form is preferable. This is obtained by cutting tubers to pulp which is dried by air or inert gas to prevent possible oxidation.

The drying process of JA slices was monitored in a laboratory drying chamber with a diameter of 80mm and a grate with free area of 6%. The slices of JA were chips 3mm in diameter, each of a different length. The drying rate was defined as a relative loss of humidity for a specific period of time. Relation for the drying rate can be defined as:

$$R = {}^{(moisture\ loss)}/_{(sample\ weight * time)} \tag{2}$$

The loss of humidity was specified gravimetrically. Drying in a stabilised state is characterised by enthalpy balance, which can be expressed as follows:

$$\dot{V} * \rho_g * c_{p,g} * (T_{in} - T_{out}) = \Delta H^0 * R * G_0 + \dot{Q}_z \tag{3}$$

The drying rate of the slices was measured at the input air temperature of 60°C. The measured data implied that at the beginning of the drying process, humidity of the material dropped linearly in time and did not depend on the size of dried particles. Water was poorly bound in the tubers and it was very simple to eliminate it. The curve for loss of humidity of slices of various sizes is shown in Figure 10.

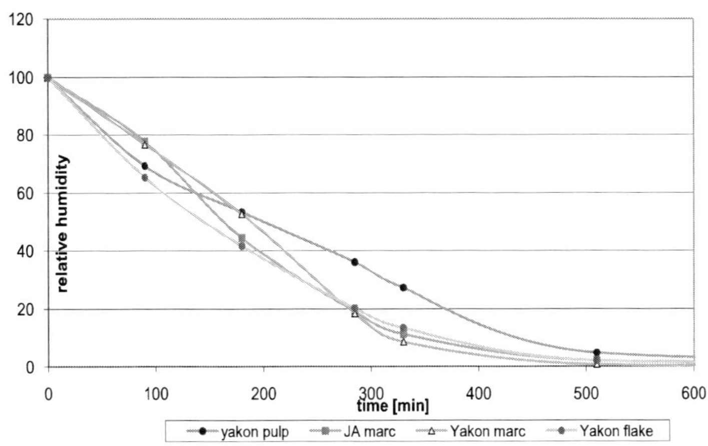

Figure 10. Drying of chips of yacón and JA. Temperature of air 60°C.

It is apparent that the drying process has two phases, in which the humidity of the material changes. In the initial phase, the loss of humidity is linear in time. In the second phase, diffusion starts to affect the humidity of the material, which plummets exponentially. The drying rate in the first phase has been defined. Transition between the first and second phase was observed at 10- 15% relative humidity for both yacón and JA.

Humidity loss in JA particles in dependence on time was measured and based on the initial drying phase. The slope of the drying rate curve was determined according to the relation (9). For loss of humidity observed for 438g of slices and flow rate of drying air 15 Nm^3/hour, see Figure 11. The inlet drying air temperature was 40°C.

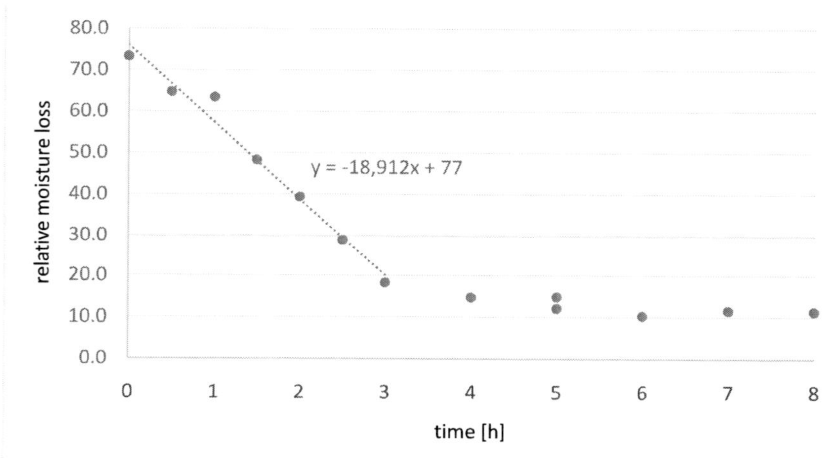

Figure 11. Drying rate.

The results imply that the relative humidity of the slices drops in time to the level of critical humidity, and therefore that the initial drying rate is constant. When humidity reaches 10wt%, the drying rate plummets dramatically in the second phase of drying. This is a threshold level for natural drying, when the tubers dry freely at an ambient temperature. The drying time may extend to several weeks.

The effect of air temperature on the kinetics of drying of pulp was studied. The balance equation may be expressed as follows:

$$(T_{in} - T_{out}) = \Delta H^{0} * \theta + \dot{Q}_z / (V * \rho * c_p) \tag{4}$$

Figure 12 shows the linear curve of dT depending on drying rate for JA tuber slices.

The slope of the regressive straight line corresponds to the enthalpy change observed during the drying process, i.e., 2.66 MJ/kg. The heat of evaporation of water is only 2.47 MJ/kg. The difference between these values corresponds to the energy consumed by the transport of water in a particle and transfer of water from liquid solution containing polysaccharides (inulin etc.) to the gaseous phase.

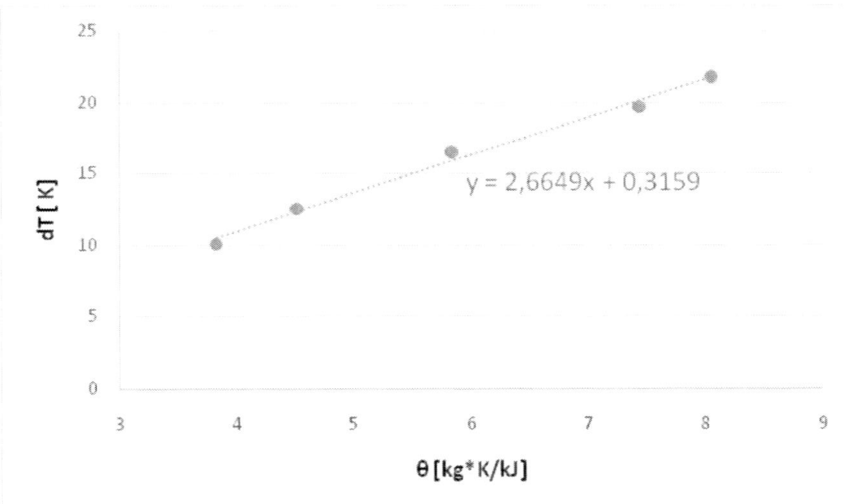

Figure 12. Drying rate in dependence on air temperature.

It was observed that when the temperature of dried pulp does not exceed 30°C, the dry matter is light in colour. With increasing the resulting product became darker. Oxidation and saccharide destruction were reported. Dried Jerusalem artichoke slices were leached in hot water in the Soxhlet extractor and the leach analysed for content of inulin, glucose, fructose, sucrose and hydroxymethylfurfural (HMF). The results are given in Table 4.

The table shows that when JA was dried at temperatures exceeding 50°C, inulin decomposed and the content of monosaccharides and HMF increased. At the same time, it was observed that dried crushed tubers turned darker, even dark brown, at temperatures above 50°C; they caramelised. It was therefore necessary to ensure that the JA pulp was dried at mild temperatures (up to 30°C), which was the method whereby dry matter of light colour with a high inulin content could have been obtained. An important characteristic of the pulp from JA tubers is that they were gluten-free.

Table 4. Content of substances in the extract in dependence on drying temperature

Temperature	Fructose	Glucose	Sucrose	Inulin	HMF
°C	% w	%w	%w	%w	mg/kg
18	0.4	0.15	12.3	56.4	0.8
24	0.6	0	12.5	55.8	1.1
30	0.6	0	12.5	77.1	1.2
38	2.2	<0.05	12.7	55.2	3.3
55	2.4	0..12	13.1	55.6	7.9
78	3.3	0.26	15.7	49.4	22.4
150*	0.2	0.08	8.2	32.2	41.6
220*	0	0	1.5	11.3	18.8

Concentration of HMF was lower than 20mg/kg of dry matter. Dried JA tubers could be then ground into chips suitable for use in bakeries as an additive (5-15%) to preserve softness of bread. Caramelisation was prevented by drying at low temperatures.

In the next step, JA tubers and chicory roots were dried at higher temperatures (250-350°C), at which significant change of colour and caramelisation took place. In this case, again the quantity of HMF was monitored. Dried pieces of tubers were then leached by water and the extract obtained used in powder form as a significant additive to coffee substitute. This substitute is produced and available in markets in the Czech Republic and elsewhere.

The effect of oxygen present in the drying media was also studied in the drying experiments conducted, in which drying was carried out using air and nitrogen. Drying at higher temperatures changed not only the colour but also aroma of the resulting dry matter. In general, the Maillard reactions were concerned with caramelisation of sugar. The effect of antioxidants on colour changes in the dry matter was also studied. Na_2SO_3 and salt of ascorbic acid (potassium ascorbate) were tested as antioxidants.

2.3. Leaching of Tubers

Washed tubers were cut into slices 3mm thick and 1kg of slices put in a mixing jar. Water was added (2.5kg) at a specific temperature (40°C, 100°C) and the suspension was mixed for 30 minutes. Liquid was separated by filtering the suspension through a Nutsche Filter. The pulp was leached again under the same conditions. The procedure was repeated and the quantity of saccharides and inulin in the filtrate determined. The scheme of this leaching procedure is shown in Figure 13.

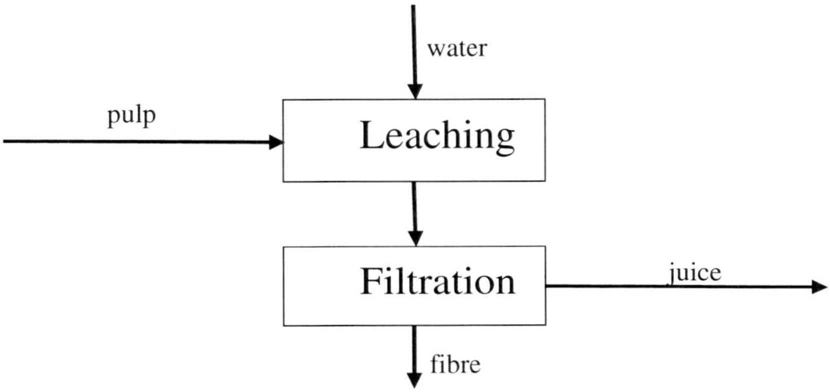

Figure 13. The leaching procedure.

Analysis of the filtrates indicated how many leaching cycles were necessary under given conditions to achieve the required mass of inulin leached out of the tubers. For the purposes of the experiment tubers of yacón and JA were leached. The results of yacón leaching are given in Table 5.

Table 5. Impact of temperature and number of leaching cycles on content of monitored substances

Juice	inulin [g/l]		sucrose [g/l]		fructose [g/l]		glucose [g/l]	
No.	40°C	100°C	40°C	100°C	40°C	100°C	40°C	100°C
F1	28.7	27.8	1.83	1.6	4.17	11.1	3.61	2.2
F2	9.43	7.85	0.69	0.37	2.71	2.66	1.06	0.5
F3	4.47	1.17	0.41	0.13	1.13	0.58	0.43	0.13
F4	1.92	0.33	0.12	0	0.57	0	0.18	0
F5	0.83	0	0	0	0.13	0	0	0

The results given in Table 5 implied that the temperature of leaching was an important parameter. This result led to the conclusion that four cycles of slices leaching was sufficient. It was also recommended that leaching should be conducted at elevated temperature, e.g., 80°C. The impact of the shape and size of slices was also monitored, and the results showed that the impact was not significant. The extract of yacón tubers contained a large mass of saccharides. Obviously the yacón tubers were very ripe and harvested a long time previously.

Slices of JA were leached under similar conditions. The main difference between the infusion from yacón and JA was that the JA infusion did not contain hexoses, just sucrose. The impact of temperature on leaching was proven.

Table 6. The impact of temperature and leaching cycles on the content of monitored substances

Juice	Inulin [g/l]		Sucrose [g/l]		Dry material [g/l]	
No.	40°C	100°C	40°C	100°C	40°C	100°C
F1	34.3	43.8	12.1	8.07	7.3	5.54
F2	9.82	10.6	4.71	1.97	2.95	1.43
F3	4.42	2.57	1.66	0.6	1.11	0.32
F4	2.37	0.6	0.62	0.11	0.4	0.08
F5	0.82	0	0.21	0	0.13	0

Another important aspect of the leaching technology was the time necessary for achieving balance between the leached material and the infusion. Therefore, the impact of leaching time on inulin concentration within one cycle was monitored. The material was leached in water at a temperature of 80°C and the weight ratio of water and slices of fresh JA was 2.5. Once mixed, the temperature dropped, therefore the suspension was heated. The required temperature was achieved after 20 minutes, so the leaching time had to be critically assessed; thus time was measured from the point at which the temperature of 80°C was reached. Experimental results can be seen in Figure 14.

The time needed for leaching JA slices was a minimum of 30 minutes, the recommended time being 60 minutes. It has been suggested [8] that the optimum weight proportion of dry tuber slices and water is 0.2. This state was achieved after 90 minutes and at 80°C. The literature refers to various data, such as a higher proportion, a lower temperature of 50°C and extraction time of up to 5.5 hours. Leaching conditions obviously depend on variety, harvest season, and the residual humidity of the slices. The leaching time for fresh slices was shorter than that necessary for dry slices or dry marc.

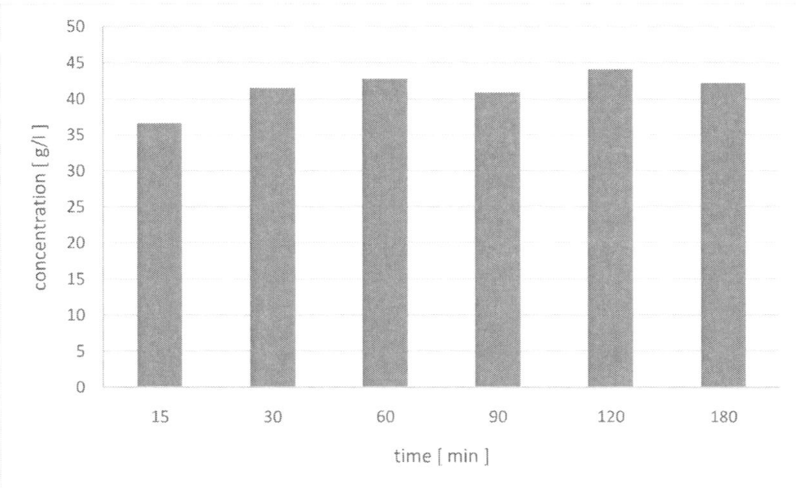

Figure 14. Influence of leaching time on concentration of inulin. T=80°C.

When leaching JA tubers harvested in November, the resulting solution contained neither fructose nor sucrose. When the experiment was repeated with tubers harvested in April, in addition to inulin and sucrose, the infusion also contained hexoses. It may thus be assumed that, as a result of activity of enzymes, polysaccharides split into simple saccharides, glucose and fructose [9].

2.4. Tuber Pressing

To obtain juice from tubers by pressing, washed tubers must be crushed first into small pieces (slices) that may, further on, be pressed to obtain raw juice. The juice yield depends on processing conditions (pressure) and the type of press used. The membrane through which juice is pressed is also an important factor.

The yield of juice was measured on a traditional press. Fresh JA tubers were crushed and put in a cloth membrane, and juice was pressed out. The yield ranged from 25-35 wt%, depending on pressure. A two-screw press is another type of press, consisting of a metallic membrane through which juice is pressed, and the yield ranges from 55-65 wt%. In addition to a high concentration of saccharides, raw juice contains enzymes and proteins, which is why it is so unstable from a microbiological point of view and must therefore be promptly processed or pasteurised. Another option for stabilising fresh juice is freezing it at temperatures below -18°C.

The composition of fresh JA juice pasteurised at different times after harvest is shown in Table 7.

Table 7. Content of monitored substances in freshly pressed JA juice

Harvest time	inulin	sucrose	fructose	glucose
	g/l	g/l	g/l	g/l
November	170	74	0	0
December	150	76	12	3
January	140	73	22	5
February	129	72	38	9
March	126	39	46	23

The advantage of pressing can be seen in the cost effectiveness of the processing, as pressing separates up to 70% of the inulin contained in the tubers. Pressing is carried out at an ambient temperature and the extracted juice is heated. With leaching, both the tuber slices and the water necessary for leaching must be heated. The energy consumption necessary for leaching is five times that needed for pressing. Moreover, proteins are retained in the leached material and cannot be separated. With pressing, although juice must be pasteurised, subsequent filtration separates the proteins and it can then be used as a special source of vegetable amino acids.

2.5. Stabilisation of Raw Juice

Raw juice is obtained by pressing crushed pieces of tubers at temperatures below 30°C. In addition to polysaccharides, the juice contains a large proportion of proteins and surface dirt as well as a number of enzymes and contaminating yeast cells. Options for eliminating the undesirable effects of these contaminants include pasteurisation or sterilisation. Which one of these is chosen depends on virulence of contaminants as well as on the following processing technology.

With pasteurisation, raw juice is heated to 80°C for 30 minutes. High temperature pasteurisation (HTP) can be used, with heating time reduced to a few seconds.

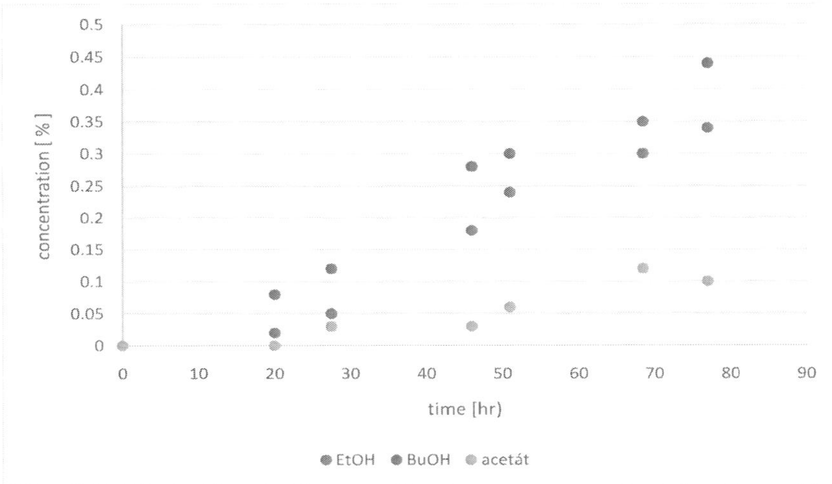

Figure 15. Stability of pasteurised yacón juice. Temperature 20°C.

Stability of the pasteurised raw yacón and JA juice was observed at 5°C and 20 °C. Juice was heated to 80°C for 30 minutes and then poured into two two-litre glass jars, one of which was placed in a fridge, the other left at the ambient temperature of 20°C. A sample of both juices was taken and the content of alcohols (ethanol and butanol) and acetate was measured in dependence on the storage time. For the results see Figure 15.

Observation of the stability of JA juice found that no alcohols were present in the sample after ten hours. Therefore, this time was considered suitable for storing of pasteurised juice at ambient temperature. The result applies not only to JA juice but also to yacón juice.

At the same time, stability of the juice stored at 5°C was observed. JA juice pasteurised at 80°C for 30 minutes was stored in an open jar in a fridge at 5°C and at an ambient temperature. The results are summarised in Figure 16.

These experiments suggest that the temperature at which pasteurised juice is stored has a significant effect on its stability and quality after pasteurisation. The results also imply that long-term stability of the juice is impossible, and if the juice is yet to be heat-treated the applicability of this protection method is somewhat limited. Clearly for industrial purposes it is essential to increase the stability of the raw juice, and therefore its sterilisation was tested at boiling for 30 minutes. The sample of sterilised juice was put in a PET bottle equipped with a water cap and the evolution of gas was monitored. The results showed that the juice sterilised at ambient temperature was stable for a period of 12 days, and when stored at temperature of 5°C, sterilised juice was stable for more than 65 days. The stabilisation method must be chosen according to technological constraints or intended usage of the product. The method using gas evolution proved to be simple and feasible even if laboratory equipment was limited.

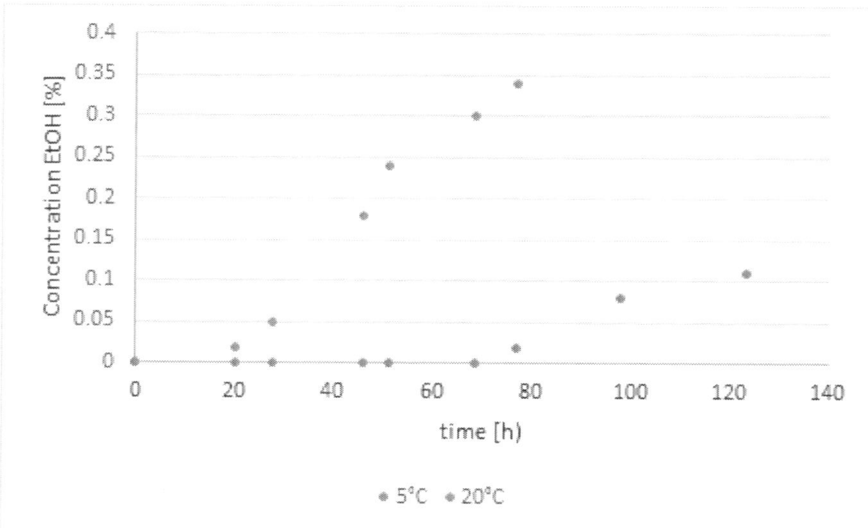

Figure 16. Dependence of the storing temperature on the juice stability.

Protein-based substances coagulate when tubers or juice are heat-treated at temperatures exceeding 50°C. If it is a raw juice, these substances coagulate together with impurities in the form of a natural pigment, into polysaccharides, including cellulose. If tubers are leached in water at a temperature higher than 60°C, these substances are not leached and are retained in the leached material as undissolved matter. Only soluble substances enter into solution, including mineral substances and organic compounds, such as glucose, sucrose, fructose and inulin.

Generally speaking, a soluble substance can be obtained from tubers by leaching or pressing. As far as leaching is concerned, a dissolving agent (usually hot water) is added to tuber slices. Pressing consists of crushing and pressing of tubers. Since tubers contain more than 65wt% of water, the pressing method is feasible and effective. Juice usually contains 15-20% of the dry mass, mostly polysaccharides. Its composition depends on the harvest season and method of storage. If tubers are to be processed to a product with a high content of inulin it is advisable to process them immediately after their vegetation period. If, on the other hand, a product with a high content of fructose is to be produced, it is advisable to process tubers harvested in spring (March or April).

2.6. Fermentation of Slices to Lactic Acid

Lactic acid is one of the most common acidulants used in the food industry. It forms as a result of fermentation in many foods, including yogurts, lacto-fermented foods, cheeses, fermented vegetables, matured meat and processed meat products.

Natural juice from tubers containing inulin also contains a varying mass of sucrose, fructose or glucose. Lactobacteria ferment the saccharides present into lactic acid under anaerobic conditions according to the equation:

$$C_{12}H_{22}O_{11} + H_2O \rightarrow 4\ CH_3CHOHCOOH \text{ , or} \tag{5}$$

$$C_6H_{12}O_6 \rightarrow 2\ CH_3CHOHCOOH \tag{6}$$

Fermentation takes place at pH 4.5 to 6 using lactobacteria. Inulin hydrolyses at pH < 5, but lactobacteria cannot process it in the same way as starch or dextrose. Hydrolysis is therefore a necessary step before fermentation to lactic acid can occur. Unlike alcohol fermentation, most types of lactobacteria process all common sugars—hexoses, pentoses, mono-saccharides, disaccharides and multivalent alcohols such as glycerol, mannitol etc. The process of fermentation ceases during production of lactic acid when the content of acid exceeds the concentration of approx. 10 g/l, i.e., 1% solution. The pH value drops to 4 and the growth of bacteria stops. [10] The process of bacterial evolution can be restarted by adjusting the pH, by adding lime, for instance.

Ground tubers of washed JA were fermented at an ambient temperature in an anaerobic environment. Ground tubers (20kg) were processed with the same mass of water. The fermenting vessel was equipped with a water cap. In this phase, so that the conditions are favourable, all microorganisms –

lactobacteria as well as *Escherichia coli* bacteria – reproduce. The infusion dimmed and foam started to evolve on the surface. A minor evolution of CO_2 was observed, which ceased within 24 hours. Residual material was separated after 120 hours of fermentation and filtrate was analysed. Lactic acid content was assessed using the HPLC method. The concentration was 1.4 wt%. The remaining inulin was found in the filtrate. Its concentration was 3.2 wt%.

3. Processing of Pasteurised Juice

Abstract

Sterile juice containing dissolved sugars and polysaccharides and filtration cake was obtained at sterilisation of raw juice and following crystallisation. The cake contained a precipitate with high content of vegetable proteins. Hydrolysis by hydrochloric acid was conducted at two temperatures. The content of individual amino acids in the hydrolysate was determined. Tryptophan was not present in hydrolysate. Acidic hydrolysis of inulin was studied and a model describing dependence of the first order rate constant on temperature and pH of solution was proposed. The effect of heterogeneous ion exchange on hydrolysis of inulin was determined. On the basis of the results obtained, a method of electrolytic separation followed by hydrolysis of natural juice from tubers of yacón and JA was developed.

Fermentation experiments of hydrolysed juice of JA to ethyl alcohol were conducted.

3.1. Introduction

At pasteurisation or sterilisation of juice, two products are obtained: protein concentrate and pasteurised juice. Both of these products can be used for further processing. Experimental verification procedures are shown in Figure 17.

Protein concentrate can be decomposed by hydrolysis into amino acids which may be further processed. Pasteurised juice contains inulin and is used as a dietary supplement. It can also be hydrolysed to fructose and used in food or as a raw material for further processing.

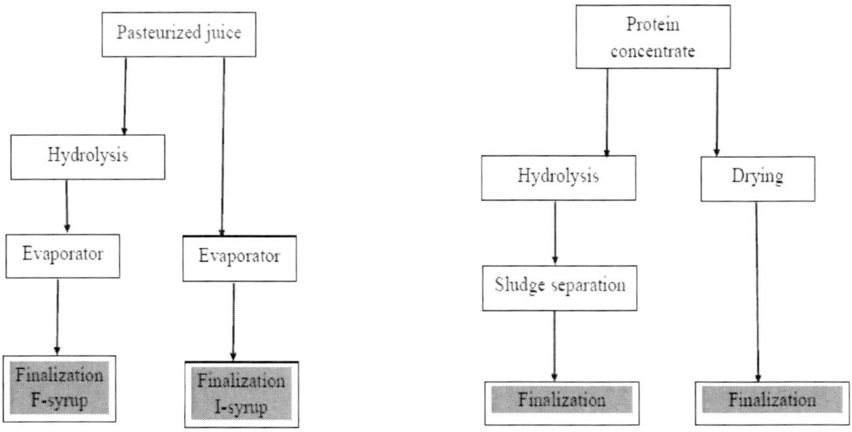

Figure 17. Schematic for processing pasteurised juices and protein concentrate.

3.2. Hydrolysis of Protein Concentrate

The origin and methods of separation of the protein concentrate were described in the preceding chapter. At leaching or pressing of tubers at temperatures lower than 50°C, juice containing sugars, enzymes, pectin and proteins was obtained. For microbial stabilisation the solution had to be heated. At heating of the fresh juice to temperatures higher than 65°C proteins precipitated (coagulated) together with other pectin-type compounds. Pasteurisation or sterilisation took place, according to the temperature used. The protein concentrate (precipitate) formed was then isolated by filtration where concentration of proteins was affected by the amount of mother liquor adhering to the filtration cake (mother liquor with dissolved sugars). But it would have been possible to wash the filtration cake and obtain concentrated protein.

A sample of protein concentrate was obtained through sterilisation of JA juice after pressing of tubers followed by filtration of the sterilised suspension. The filtration cake was not washed, and contained 73% water. It was dried in a laboratory drier at 60°C, and then disintegrated on a fragmentation mill. The main protein components were determined using elementary analysis. The results are given in Table 9.

Protein concentrate obtained thus was used as a raw material for acidic hydrolysis. The obtained product was a mixture of amino acids which formed the original protein. The usual catalyst was a mineral acid (hydrochloric, sulphuric or phosphoric acid) of 5N to 7N concentration. Recommended time for hydrolysis was from 30 minutes up to 50 hrs. [12] Favourable conditions depended on the type of the original protein and had to be determined experimentally.

Table 9. Elementary analysis of protein

N	S	C	H
% w.	% w.	% w.	% w.
2,94	0,18	45,9	6,36

It was possible to study the hydrolysis of protein components using Kjeldahl's analytical method for nitrogen. If the composition is known for each amino acid, nitrogen content makes it possible to establish the composition of these amino acids and subsequently proteins from JA. Selected amino acids are given in Table 10.

Table 10. Major amino acids in product after hydrolysis and their composition

Amino acid	Formula	MW	composition	composition	composition	composition	composition
			N	S	C	H	O
Aspartic acid	C4H7NO4	133,11	0,105176		0,360604	0,052588	0,481632
Alanin	C3H7NO2	89,09	0,157144		0,404086	0,078572	0,360198
Serin	C3H7NO3	105,09	0,133219		0,342564	0,06661	0,457608
Lysin	C6H14N2O2	146,19	0,191532		0,49251	0,095766	0,220193
Glutamic acid	C5H9NO4	147,13	0,095154		0,407803	0,06117	0,435873
Leucin	C6H13NO2	131,18	0,106724		0,548864	0,0991	0,245312
Glycin	C2H5NO2	75,07	0,186493		0,319702	0,066605	0,427201
Methionin	C5H11NO2S	149,21	0,093827	0,214463	0,402118	0,073722	0,21587
Average			0,133659	0,214463	0,409781	0,074267	0,355486

The calculated average composition was the base for determining the content of protein in the dry matter. According to the calculation, the protein concentrate contained approximately 20-25 mass% of proteins; the remainder, approximately 20 mass%, was a mixture of carbohydrates that passed through the sieve at pressing.

Conditions of hydrolysis of the protein concentrate obtained at sterilisation of raw juice from JA tubers were experimentally verified. The time necessary for reaching a high conversion of nitrogen compounds soluble in water at boiling point at atmospheric pressure was measured. Hydrochloric acid concentration was selected as a parameter. The results of these experiments are presented in Figure 18.

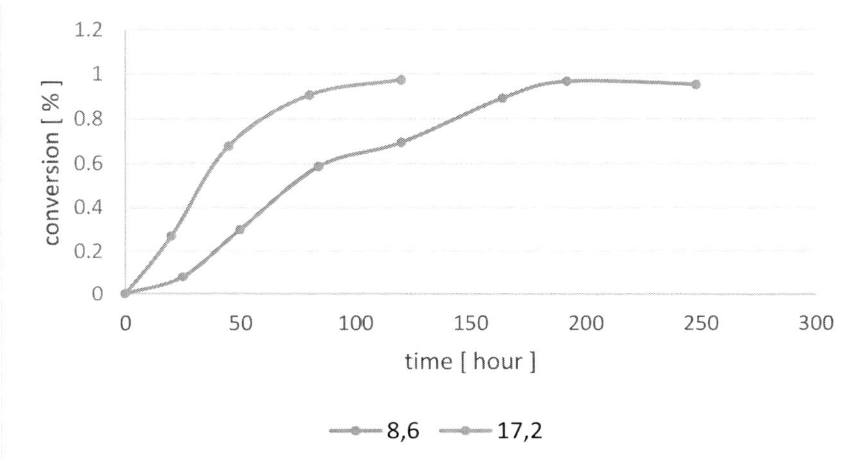

Figure 18. Protein hydrolysis at 100°C. The effect of HCl concentration.

It is clear from these experiments that the rate of hydrolysis is dependent on the concentration of acid in the solution. The time needed to reach a high conversion at 100°C and atmospheric pressure is several hours. The relatively short time of hydrolysis and its conditions indicate that the vegetable protein of JA is composed of a small number of units, and thus that it can be considered to be of the peptide type.

In the next step, the effect of temperature on the rate of hydrolysis was studied. Hydrolysis of protein concentrate was carried out in an autoclave at a temperature of 140°C and concentration of hydrochloric acid 8.6 mass%. The autoclave was of 2.5l volume and had a magnetic mixer and temperature and pressure measurement. The autoclave was heated by external electric heater. Hydrolysis was carried out by heating the charge in the autoclave and then kept at the required temperature for 60 minutes. The experiment was then terminated by switching off the electric heating, and the batch was left to cool. The results of these measurements are presented in Graph 19.

It is clear from this graph that at higher temperature of 145°C the maximum conversion is reached after 90 minutes inclusive heating of the autoclave. The same conversion is reached at 100°C only after 180 minutes. The rate of hydrolysis is strongly affected by temperature and acid concentration. But it must be noted that under conditions of acidic hydrolysis,

hydrolysis of the polysaccharides present also takes place where the formed products contaminate the mixture of amino acids.

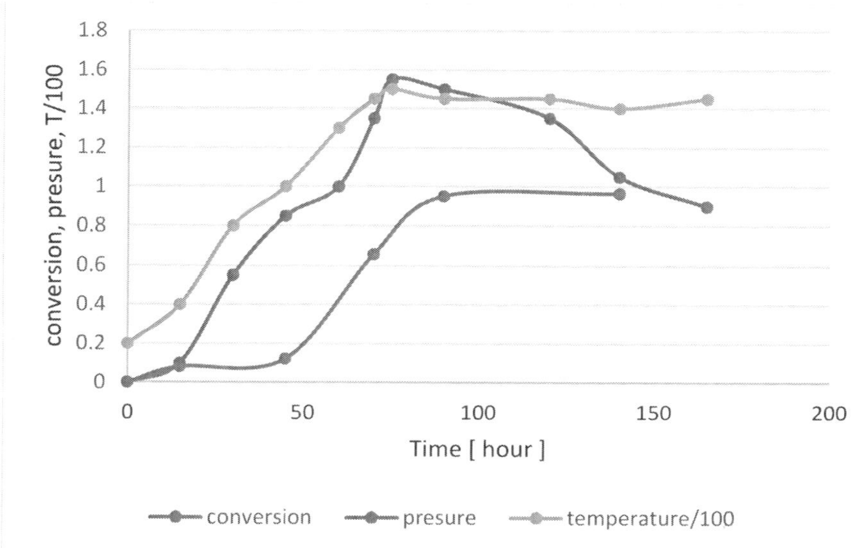

Figure 19. Hydrolysis of protein by hydrochloric acid under pressure.

3.2.1. Valuation of Protein in Jerusalem Artichoke

Vegetable proteins are a valuable nutritional component and their quality can be characterised by the quantity of essential acids they contain. The composition of protein amino acids from vegetable protein concentrate of JA is given in the next Table. The content of essential amino acids is nearly 50%. From this point of view, the protein present in JA is of a good duality, but it does not contain tryptophan, one of essential amino acids important in nutrition.

Table 11. Composition of protein amino acids from vegetable protein concentrate of JA

Asp	Thr	Ser	Glu	Gly	Ala	Val	Met	Ile	Leu	Tyr	Phe	His	Lys	Arg	Trp
% w	% w	% w	% w	% w	% w	% w	% w	% w	% w	% w	% w	% w	% w	% w	% w
10,5	6,1	6,3	11,3	10,2	8,9	8	1,7	5,7	8,5	3,4	5	2,2	7	5,2	0

3.3. Processing of Juice

The raw juice of the tubers after pressing and subsequent sterilisation, or the extract after leaching of the tubers by hot water, contains, as well as carbohydrates, primarily inulin. It contains neither proteins nor pectins. The quantity and composition of carbohydrates in this juice depends on the method of processing, the origin of the tubers and the time of their harvesting. This juice can be concentrated to inulin syrup (I-syrup) and can be used as a food additive, as well as for the separation of pure inulin.

Another procedure is based on hydrolysis of inulin when juice with high fructose content is obtained. Such solution could be also concentrated where fructose syrup is obtained. Alternatively, it is possible to utilise the hydrolysate for fermentation where ethyl alcohol, lactic and/or citric acids are obtained.

Inulin-inclusive oligofructans are not digested by the human organism in the upper part of small intestine, as the human digestive system does not contain the enzyme amylase, a hydrolytic enzyme necessary for the digestion of inulin. The nutritional value of inulin is based in its physiologic effect of fermentation in the lower part of gastrointestinal tract, where bacteria of the tribes *Bifidus* and *Lactobacillus* are present. Inulin is not sweet, but it can be decomposed by hydrolysis to monosaccharides, which are suitable as sweeteners.

3.3.1. Hydrolysis of Inulin

Inulin in a neutral medium is stable at temperatures up to the boiling point of the solution. In an acid medium, hydrolysis of the polymer chain to individual monosaccharides occurs. Total hydrolysis of inulin is an alternative which could be applied for production of high quality fructose. Such a process can be carried out either by fermentation or by a chemical process. Hydrolysis controlled chemically is cheaper and, as far as process conditions are concerned, less demanding. Chemical hydrolysis can be easily controlled by the quantity of catalyst used. As a homogeneous catalyst, sulphuric hydrochloric, phosphoric, citric, oxalic and tartaric acid are utilised. [13] Hydrochloric acid is considered the most suitable catalyst for hydrolysis of inulin. In hydrolysis with a heterogeneous catalyst, instead of acids, a cation exchanger is used. A solid ion exchanger can be easily separated from the reaction product after hydrolysis. [14]

At hydrolysis the chain of inulin is broken into shorter chains through a step-by-step process. Oligofructans and inulin with a lower molecular weight form. An inulin molecule could break into a molecule of fructose and molecule of inulin, which has a chain just one molecule of fructose shorter. Hydrolysis continues by breakage of chains to shorter molecules up to the final stage of molecules of fructose and glucose. Sacharose also appears during hydrolysis; this is one of the intermediate stages of the reaction. The whole process takes place in an acid medium and can be described by the following scheme:

Inulin → FOS, inulin, sucrose, glucose, fructose → **glucose, fructose**

Inulin is predominantly a mixture of linear β-(1-2)-linked fructose chains with a terminal glucopyranose unit. There can be a small percentage of inulin chains β-(2-6)-linkages. In addition, a small percentage of inulin molecules do not contain a terminal glucopyranose unit. These have a terminal fructoside unit [15,16].

Inulin is often characterised by the degree of polymerisation (DP) as the ratio of molecules of fructose and glucose forming its molecule. If DP<10 then the terms fructooligosacharides (FOSs) and/or oligofructans are used. In the literature [16] it is claimed that the DP of inulin could be in the range of 2-70. The process for the splitting of inulin with long chain (DP>20) becomes ultra- or nano-filtration. The ratio of inulin with short chain is processed in the food and meat industries as an additive or used for production of pure fructose.

3.3.1.1. Heterogeneous Catalysis

Fructose as a sweetener is nearly as twice as sweet as the most common sweetener, sucrose. It is therefore often used as a substitute for sucrose, and it is claimed that the calorific value of soft drinks is by this substitution significantly reduced. One of the progressive raw materials for production of fructose is inulin. [12,16] To obtain fructose, inulin must be processed by hydrolysis. This could be executed either by an enzymatic or chemical process. Chemical hydrolysis is carried out in an acid medium with a homogeneous or heterogeneous catalyst.

The basic principle of heterogeneous catalytic hydrolysis could be divided into following process steps. The first step is the preparation of an aqueous solution such as juice or aqueous solution containing inulin. The second step is the ion exchange at higher temperature (40-70°C) on acid ion exchanger in the H^+ cycle. The third step is the hydrolysis of inulin itself at temperatures of 70-100°C. This is followed by the second ion exchange, in which the solution is deprived of acid anions using an anion exchanger, when the pH of the solution is adjusted to 6.5-7. Then the solution is thickened to the required concentration of fructose syrup. The process variants are specified in patents [17-19]. One alternative to classic ion exchangers is the use of zeolites [16]. There are basic differences between use of ion exchangers and zeolites. Ion exchangers have a capacity of 4-5 meq/g and specific surface area of 2-10 m^2/g. Zeolites have an ion exchange capacity of 1-1.5 meq/g and a specific surface area of 400-500 m^2/g. The size of pores influences the hydrolysis of inulin. The size of the catalyst surface area is manifested by adsorption of fructose.

Heterogeneous hydrolysis of inulin at 100°C with the cation exchanger OSTION in H^+ cycle was studied. Graph 20 presents the dependence of inulin concentration on time for different concentrations of catalyst.

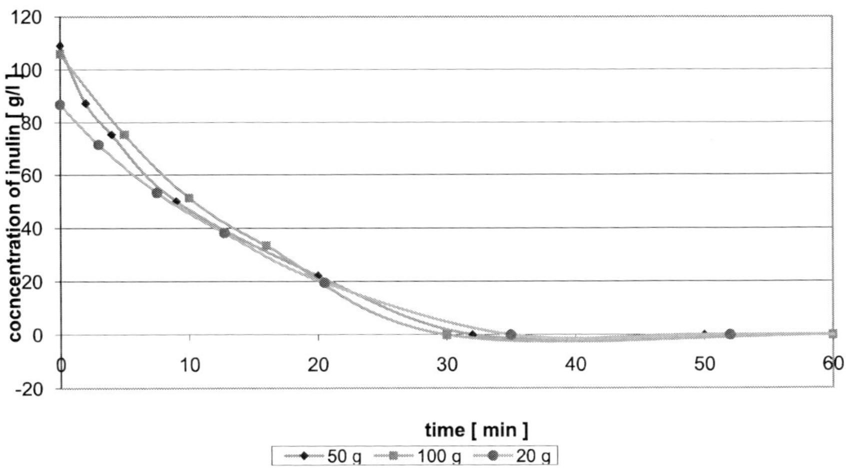

Figure 20. Change in concentration of inulin at temperature 100°C. Catalysis by ion exchanger.

The rate constant of hydrolysis of inulin at 100°C is independent of the quantity of catalyst. According to decrease of concentration it could be assumed that the hydrolysis is controlled by the first order kinetics. The results obtained are given in Table 12.

Table 12. Rate constants for different concentration of the ion exchanger at 100°C

Cation exchanger	k	pH
g/l	1/min	
20	0,0825	2,4
50	0,0869	2,8
70	0,0808	2,2
100	0,0855	2,5

It is clear from this table that the addition of the cation exchanger (OSTION) to the solution containing inulin changes the reaction medium in such a way that pH becomes lower than 3. This result is in agreement with the literature [19]. The use of heterogeneous catalyst is advantageous for its easy separation when hydrolysis is terminated, while pH remains the same.

However, it was recorded that the concentration of the fructose formed changed at hydrolysis of inulin in dependence on the quantity of ion exchanger used. It is therefore possible to assume that at the boiling point, a portion of monosaccharides is adsorbed on the surface of cation exchanger. Dependence of fructose concentration on time for different quantities of cation exchanger is given in Graph 21.

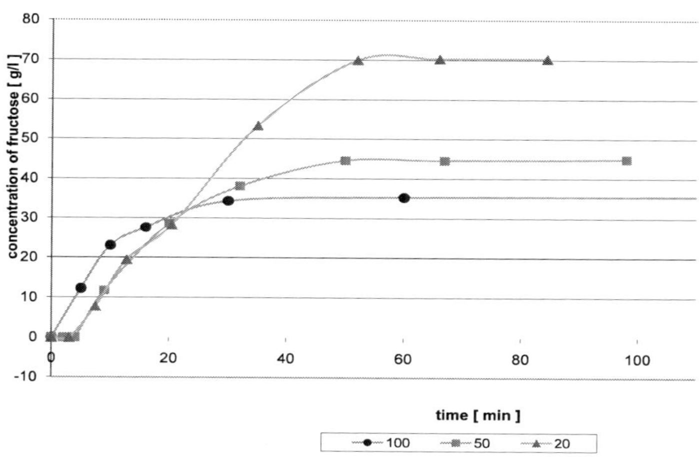

Figure 21. Concentration of fructose in dependence on time for different quantities of cation exchanger.

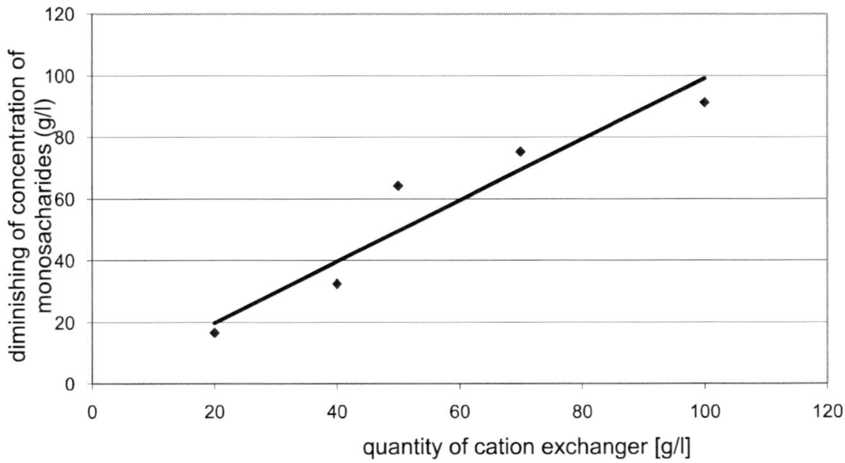

Figure 22. Dependence of diminishing concentration of monosaccharides on quantity of cation exchanger.

It is therefore probable that inulin contains some percentage of mineral salts whose cations are by cation exchanger exchanged to H^+ ions so that the pH of the solution is reduced. In solution cations are adsorbed by the ion exchanger so that the pH of the solution is reduced and hydrolysis takes place. It could be concluded that the cation exchanger in this case is not acting as a catalyst but as a simple ion exchanger.

It is clear that the cation exchanger used acts as a sorbent. This sorption is the reason for the change in colour of the catalyst. The original colour of the catalyst is similar to honey, turning to black after utilisation; see Figure 23.

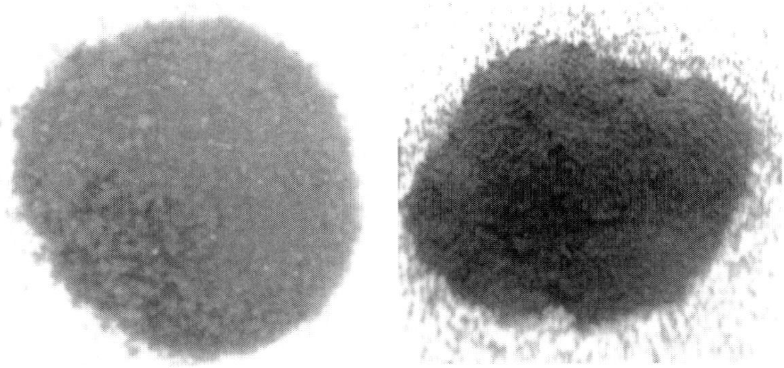

Figure 23. Comparison of fresh and utilised catalyst, showing change in colour.

This change of colour does not have an effect on the ion exchange activity of the cation exchanger. The cation exchanger can be used again as a catalyst. It is only the ion exchange capacity that changes. With decreasing ion exchange capacity the quantity of adsorbed cations that is accompanied with increase of pH also decreases. This affects the rate of hydrolysis, which is the reason why the pH of juice subjected to hydrolysis must be followed.

At temperatures above 100°C in an acid medium the effect of the competing reaction (dehydration of fructose) at which 5-hydroxy-methylfurfural (HMF) forms increases. But HMF is not adsorbed on the catalyst and passes into the solution where it contaminates the prepared fructose syrup. HMF is suspected of cytotoxic, mutagenic, carcinogenic and genotoxic effects on the human organism. The Codex Alimentarius of the WHO and European Union determines the maximum permissible quantity of HMF in food [25]. HMF does not form at hydrolysis of inulin using a cation exchanger at temperature of 100°C.

3.3.1.2. Homogeneous Catalysis

Hydrolysis of inulin is a reaction dependent on pH [33-36]. Consistent with the literature [38], in neutral and alkalic media the reaction does not take place. The advantage of chemical hydrolysis lies in the simplicity of its execution and the ease by which the reaction can be controlled through control of pH and reaction temperature [37]. At high temperature or in a very acidic medium, decomposition of fructose occurs [36]. Inulin is a polysaccharide and its hydrolysis is dependent on the degree of polymerisation, DP. It has been verified experimentally that hydrolysis of inulin dissolved in water does not take place at 100°C at pH > 4,5.

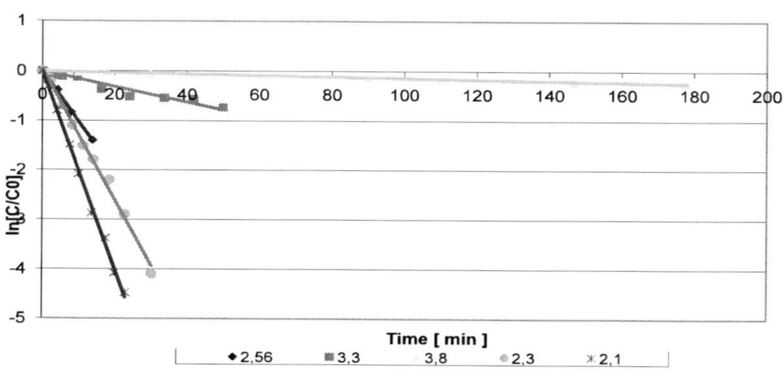

Figure 24. Determination of rate constants for the first order reaction. Temperature 100°C.

Dependence of the rate constant of hydrolysis of inulin (DP<10) on pH at a constant temperature of 100°C (boiling point of the mixture) was studied. As a homogeneous catalyst, sulphuric acid was used. The results obtained are presented in Figure 24.

It is clear from this graph that the rate of hydrolysis of inulin at constant temperature 100°C decreases with increasing pH, and the rate of hydrolysis could be described by the 1st order kinetics. The slope of the direct lines in this graph corresponds with the reaction constant at the boiling point of the mixture.

Dependence of the rate constant on pH at 100°C is given in Figure 25.

From this analysis came the conclusion that the dependence of rate constant of hydrolysis of inulin on pH has the form:

$$\ln(k) = a + b * (pH) \tag{7}$$

By regression analysis, the constants a= 9,058 a b= -3,04 were determined.

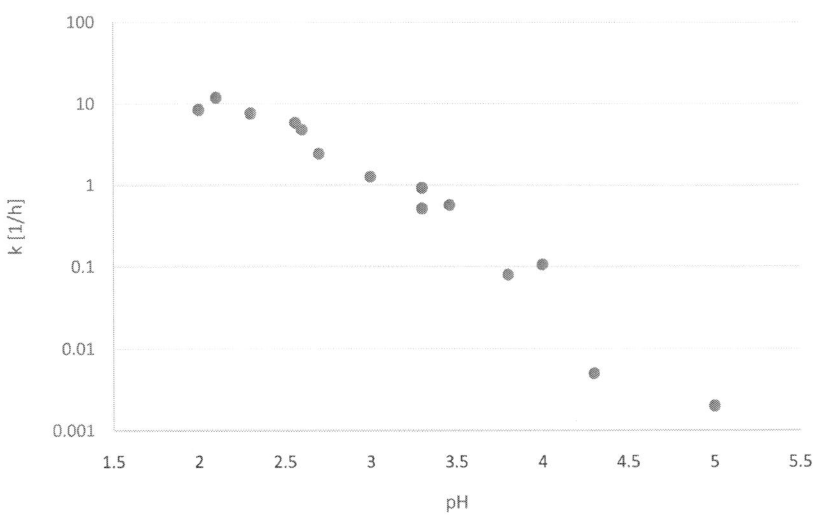

Figure 25. Dependence of rate constant on pH of solution. Temperature 100°C.

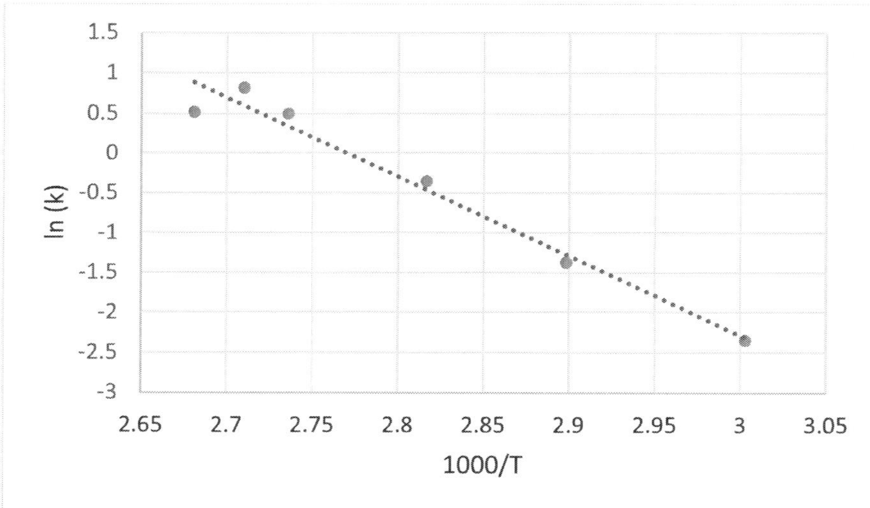

Figure 26. Dependence of rate constant on temperature. pH=3.

The rate of hydrolysis was dependent on temperature, where validity of the Arrhenius equation was assumed. The rate of hydrolysis was measured at constant pH=3 at different temperatures. In the next graph, dependence of the rate constant on temperature was given at hydrolysis of inulin catalysed by sulphuric acid.

This graph indicates that dependence of the rate constant of hydrolysis was in agreement with the Arrhenius dependence. The rate constant for the diminishing of inulin at hydrolysis catalysed by sulphuric acid then had the form:

$$k = 30{,}312 * \exp\left(-\frac{7927}{T} - 1134 * \frac{pH}{T}\right) \tag{8}$$

The validity of this equation was verified experimentally for the range $1{,}8<pH<5$ and for temperatures $330<T<375$

Correlation (7) was verified not only at hydrolysis of pure inulin but also with the natural juice of JA.

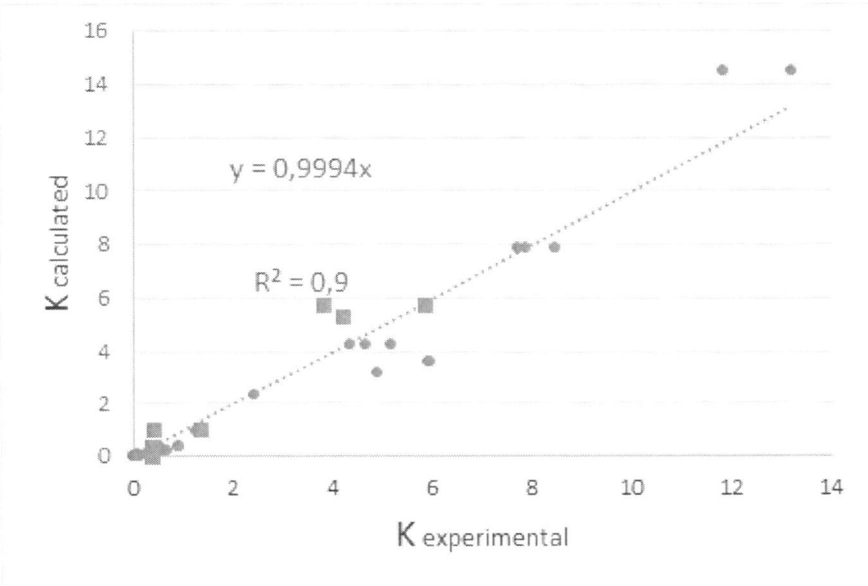

Figure 27. Comparison of calculated and experimental rate constants of hydrolysis of inulin. The squares are rate constants for natural juice containing inulin of JA. The dots are rate constants for pure inulin.

Homogeneous hydrolysis of inulin is one of methods by which to determine the degree of polymerisation DP [16, 38]. Assuming that a molecule of inulin is formed by one molecule of glucose and **n** molecules of fructose, hydrolysis of inulin can be written in the form:

$$GF_n + n * H_2O \rightarrow G + n * F \tag{9}$$

If, by an analysis, concentration of monosaccharides is determined, the degree of polymerisation DP is equal to ***n***. This analysis is affected by the number of molecules in the chain of fructose of the terminated fructoside unit in the pyranose form.

3.3.1.3. Electrolytic Hydrolysis

The final product of hydrolysis of inulin (hydrolysate) carried out through heterogeneous or homogeneous catalysis always contains a certain quantity of acids. Hydrolysate usually has a pH close to 2. This result is reached through use of ion exchangers or by addition of acids. For further processing or use in the food industry this acidity is a problem which needs to be resolved: acids must be neutralised. This can be done again by the use of ion exchangers or though the addition of an alkali. Where the former method is used, the fructose syrup is deprived of natural mineral compounds; in the latter method the syrup is enriched for the salt of the respective acid.

A method of separation of cations from natural juice has been proposed whereby juice is filled into the anode chamber of an electrolyzer in which the chambers are separated by a cation exchanger membrane and due to passage of electric current cations are separated. By this step pH in the anode chamber is reduced to round 2-3. The acid solution formed is transferred into a heated collection (storage) tank in which hydrolysis takes place for the required time. Hydrolysate is then filled into the cathode chamber into which pass the missing anions by which step adjustment of pH to original value takes place. The formed hydrolysate is thickened to the required concentration e.g., by evaporation.

One advantage of this procedure is that no chemical compounds are either added to or removed from the natural solution of inulin. Thus, a bioproduct is produced. Another advantage is that, by control of electric current, temperature and/or residence time, it is possible to control the degree of hydrolysis by which a total solution of monosaccharides could be prepared, but also a partial hydrolysate of inulin, which would be a mixture of inulin molecules with different degrees of polymerisation and content of fructose molecules by which the properties of the inulin syrup could be adjusted.

The described process and unit represent one possible application of an invention. This is protected by a patent [31]. The authors hope that it is clear that the described invention could be applied in other arrangements.

3.4. Fermentation of Hydrolysate to Ethanol

Inulin syrup can be processed by fermentation directly to a solution containing ethyl alcohol. Fermentation must be performed by special yeast strain where conditions for its activity must be strictly observed. In industry this process is usually divided into two separate steps. First, inulin is chemically hydrolysed, and then the formed fructose solution is fermented by yeast *Saccharomyces cerevisiae* to ethyl alcohol solution with maximum concentration of ethyl alcohol within the range of 10-12 vol.%. Fermentation takes place according to equation:

$$C_6H_{12}O_6 \rightarrow 2\ C_2H_5OH + 2CO_2 \tag{10}$$

This fermentation is accompanied by formation of a gas (CO_2) which fills the reaction space and which also signalises the proceeding reaction. This process is well-developed and both industrial and commercial ethyl alcohol is produced in this way.

In theory, from a hydrolysate containing 180g of hexoses form 92g of ethyl alcohol and 88g of CO_2. It has been found that about 5% of ethyl alcohol is lost in venting gases and 3.5% of hexose is not fermented. This means that out of 100g of fructose, 46.8g of ethanol are obtained, which corresponds to 59ml of 100% alcohol and to 62ml of 95% ethyl alcohol. In the literature [40] the yield from 1kg of JA tubers is quoted as being from 65 to 80ml of 95% ethanol.

Laboratory results indicate that from 1kg of JA tubers by acid hydrolysis at 100°C, approximately 110-115g of glucose and fructose are obtained. After hydrolysis, the solution is neutralised to pH=6. This solution is then fermented by commercial food yeast at laboratory temperature. The fermentation process is controlled by the quantity of formed CO_2. After four days, fermentation is terminated, and next day the yeast is filtrated off and the filtrate distilled. By fermentation, 45g of absolute ethyl alcohol is obtained from 100g fructose. On basis of a chromatographic analysis, distillate also contains ethyl acetate, propanol, 2,3-butandiol and butanol, but no methyl alcohol.

Yield of ethyl alcohol would depend firstly on the content of saccharides and inulin in JA tubers, and also on the method of their leaching and sugar yield after hydrolysis and the subsequent fermentation process. The literature [40] cites various economically more advantageous types of JA tubers. For these reasons, comparison of results would always have a certain scatter. This process is well-developed as far as the process side is concerned, but is still the subject of research [41-46] and pending patent protection [47-50].

Notation

H_u	minimum heating value	$kJ.kg^{-1}$
m_c	carbon mass ratio in fuel	%
m_H	hydrogen mass ratio in fuel	%
m_O	oxygen mass ratio in fuel	%

m_S	sulphur mass ratio in fuel	%
m_{H2O}	water mass ratio in fuel	%
V	airflow	m^3/hr
ρ_g	density of air	kg/m^3
$c_{p,g}$	specific heat of air	kJ/(kg*K)
T_{in}	inlet temperature of air	K
T_{out}	outlet temperature	K
ΔH^0	heat of vaporisation	kJ/kg_{water}
R	rate of evaporation of water	$kg_{water}/(kg_{inp.mat}*hr)$
G_0	input material into dryer	kg
$\dot{Q}_z$	loss of heat dryer	kJ/hr

ACKNOWLEDGMENTS

The authors thank Mrs. Eva Macháčková and Mr. Jindřich Karban for all analyses and Mr. Roman Pavela for testing of extracts. Many thanks to Mr. Milos Machacek and Mr. Milan Rylek for proofreading the text. This research was funded by Grant No. TE01020080 (Competence Centre BIORAF) from the Technology Agency of the Czech Republic.

REFERENCES

[1] Barta, J., Jerusalem artichoke as a multipurpose raw material for food products of high fructose or inulin content, in *Inulin and Inulin-Containing Crops,* Fuchs, A., Ed., Elsevier, Amsterdam, 1993, pp. 323–339.

[2] Aravina, L.A., Gorodetskii, G.B., Ivanova, N.Y., Komarov, E.V., Momot, N.N., and Cherkasova, M.A., *Extraction of Inulin and Other Fructan-Containing Products from Jerusalem Artichoke and Other Inulin-Containing Raw Materials,* Russian Federation Patent 2175239, 2001.

[3] Kays, S.J., and Nottingham, S.F., *Biology and Chemistry of Jerusalem Artichoke,* CRC Press website at http://www.crcpress.com, 2008.

[4] Lyčka, Z.: Importace of peletisation of wood mass (in Czech). Biom.cz [online]. 2011-04-18 [cit. 2014-09-22]. <http://biom.cz/cz/odborne-clanky/vyznam-peletizace-drevni-hmoty>. ISSN: 1801-2655.
[5] Cagaš, B., Macháč, R., and Nedělník, J., Grassland Science in Europe: Alternative Functions of Grassland. Brno: *15th European Grassland Federation Symposium* 2009, 368-371, 2009.
[6] Stupavský, V. Pellets from biomass – wood, plant and bark pellets (in Czech) *Biom.cz* [online]. 2010-01-01 [cit. 2014-09-22]. <http://biom.cz/cz/odborne-clanky/. ISSN: 1801-2655.
[7] Čepl J., Vacek J., and Bouma J., *Process of planting and utilisation of JA* (in Czech) UZPI Praha, 9, 20-28, (1997).
[8] *Stonefly Industries*, Bryan, TX, USA.
[9] Khalil S.E., Abd El- Aziz N.G., and Abou Leila B.H.: Effect of Water Stress, Ascorbic Acid and Spraying Time on Some Morphological and Biochemical Composition of Ocimum basilicum plant, *J. Am. Sci.* 6, 36-42, (2010).
[10] Luedeking R., Piret E.L., A Kinetic study of the Lactic Fermentation. *J.Biochem.Microbiol. Technol.* Eng. Vol.1 No.4., 393-412, (1959).
[11] Ge X.Y., Qian H., and Zhang W.G. Improvement of lactic acid production from Jerusalem artichoke tubers. *Bioresour. Technol.* 100, 1872-74, (2009).
[12] Akira TSUGITA and Jean-Jacques SCHEFFLER, A Rapid Method for Acid Hydrolysis of Protein with a Mixture of Trifluoroacetic Acid and Hydrochloric Acid, *Eur. J. Biochem.* 124, 585-588 (1982).
[13] Abasaeed A.E. and Lee Y.Y., Inulin Hydrolysis to Fructose by a Novel Catalyst, *Chem. Eng. Technol.* 18, 440, (1995).
[14] U.S. patent No. 3,044,904 (1964).
[15] De Leenheer, L. and Hoebregs, H., *Progress in the elucidation of the composition of chicory inulin,* Starch/Stärke, 46, 193–196, 1994.
[16] De Leenheer, L., Production and use of inulin: industrial reality with a promising future, in *Carbohydrates as Organic Raw Materials III,* van Bekkum, H., Röper, H., and Voragen, A.L.J., Eds., Weinheim, Cambridge, U.K., 1996, pp. 67–92.
[17] Patent WO1999037686A1.
[18] U.K. patent No. 1,083,500 (1964).
[19] *Biotechnology and Bioengineering*, Vol. XX, 4. 447450 (1978), John Wiley & Sons, Inc.
[20] US patent 4,613,377.
[21] Roberfroid, M.: *Crit. Rev. Food Sci. Nutr.* 33 (2): 103,(1993)

[22] Saengthongpinit, W., *PhD Thesis, Kasetsart University, Thailand,* (2005), ISBN 974-9834-71-2.

[23] van Dam, H.E., Kieboom, A.P.G., and van Bekkum, H., *The conversion of fructose and glucose in acidic media: formation of hydroxymethylfurfural,* Starch/Stärke, 38, 95–101, 1986.

[24] Fuchs, A., *Potentials for non-food utilization of fructose and inulin,* Starch/Stärke, 39, 335–343, 1987.

[25] Alinorm: 01/25, 2001; directive EU: 2001/110/ES.

[26] Carabin, I.G. and Flamm, W.G., Evaluation of Safety of Inulin and Oligofructose as Dietary Fiber. *Regul. Toxicol. Pharmacol.* 1999, 30, 268–282.

[27] Franck, A. D.; Leenheer, I. L. *Polysaccharides, Biopolymers online,* 2005th ed.; Wiley-Blackwell, Chapter Inulin, pp 441–473.

[28] Abasaeed, A. E.; Lee, Y. Y. Inulin Hydrolysis to Fructose by a Novel Catalyst. *Chemical Engineering & Technology* 1995, 18 (6), 440–444.

[29] Saemgkanuk, A., Nuchadomrong, S., Jogloy, S., Patanothai, A., and Srijaranai, S.A., Simplified spectrophotometric method for the determination of inulin in Jerusalem artichoke (Helianthustuberosus L.) tubers. *European Food Research & Technology* 2011, 233 (4), 609–616.

[30] RU patent 4613377 Yamazaki, H. and Matsumoto, K. Production of fruktose syrup.

[31] CS patent 304803. *Method of hydrolysis of inulin solution and device for its execution (in Czech).*

[32] Kim, S. B., and Lee, Y.Y., Effect of diffusion in solid acid catalyzed inulin hydrolysis. *Appl. Biochem. Biotechnol.* 1988, 17 (1-3), 55–72.

[33] Conti, F.N., *Versuche zur gewinnung von sirup aus topinambur.,* Zucker, 6, 120, (1953).

[34] Guiraud, J.P. and Galzy, P., Inulin conversion by yeasts, *Ind. Aliment. Agric.*, 98, 45, (1981).

[35] Ohira, T. and Kobayasi, K., The preparation of sugar from Jerusalem artichokes, *J. Agr. Chem. Soc.,* 23, 103, (1949).

[36] Pekič, B., Slavica, Z. Lepojevič, and Petrovič, S.M., Effect of pH on acid hydrolysis of Jerusalem artichoke inulin., *Food Chemistry* 17, 169, (1985).

[37] Denglin L., Wei, X., Jianxue L., and Shengnan, L., Acid-induced degradation behavior of inulin. *Advances in Biomedical Engineering Vols.* 1-2, 2011.

[38] Kamada, T., Nakajima, M., Nabetani, H., Saglam, N., and Iwamoto, S., Availability of membrane technology for purifying and concentrating oligosaccharides, *Eur. Food Res. Technol.,* 214, 435–440, 2002.

[39] Luedeking, R. and Piret, E.L.: Journal of Biochemical and Microbiological Technology and Engineering, 1, 393, (1959).

[40] Heyraund A., Rinaudo M., and Taravel, F.R., Isolation and characterisation of oligosaccharides containing D-fructose from Juices of Jerusalem artichoke, kinetic constants for acid hydrolysis. *Carbohydrate Research,* 128, 311-320, (1984).

[41] Kays, S., *Nottingham: Biology and Chemistry of Jerusalem Artichoke,* CRC Press, 2008.

[42] Ohta, K, Hamada, S., and Nakamura, T., Production of high concentrations of ethanol from inulin by simultaneous saccharification and fermentation using Aspergillus niger and Saccharomyces cerevisiae, *Appl. Environ. Microbiologz,* 59, No.3,729, 1993.

[43] Manso, O.L. and Giraldo, G.A., Acid hydrolysis of inulin to obtain a fermentable juice for bioethanol production. II congreso national de technologia de los alimentos, *Cartif,* 556, (2011).

[44] Sanchez, O.J. and Cardona, C.A., Trends in biotechnological production of fuel ethanol from different feedstocks., *Bioresor. Technol.* 99, 5270, (2008).

[45] Li, S.Z., and Chan-Halbrendt, C., Ethanol production in the People's Republic of China: potential and technologies, *Appl. Energy* 86, S162, (2009).

[46] Helwani, Z., Othman, M.R., Aziz, N., Fernando, W.J.N., and Kim, J., Technologies for production of biodiesel focusing on green catalytic techniques: a review, *Fuel Process. Technol.* 90, 1502, (2009).

[47] RU patent 2144084 (2000).

[48] RU patent 2195494 (2002).

[49] GB patent 1030740 (1966).

[50] Middlehoven, W.J., Van Adrichsem P.P.L., Reij, M.W., and Koorevaar, M., Inulin-Containing Crops, Fuchs, A. (ed.), Elsevier, Amsterdam (1993).

In: Inulin
Editor: Michael G. Marlow
ISBN: 978-1-63463-985-9

Chapter 3

INULIN: MODULATION OF INTESTINAL MICROBIOTA AND PROPECTS IN NEW PRODUCTS

Marilyn Gonçalves Ferreira Kuntz, Ricardo Fernandes, Vinicius André do Rosário, Giovanna Medeiros Rataichesck Fiates and Erasmo Benicio Santos de Moraes Trindade
Federal University of Santa Catarina, Brazil

ABSTRACT

A healthy microbiota is defined as one that preserves and keeps it the wellbeing and the disease-free, especially of the gastrointestinal tract with three main functions: metabolism function, protection function and trophic role on the proliferation and differentiation of the intestinal epithelium and the development of the immune system modulation. However, some factors may influence this process of colonization, such as diet.

In recent decades, there has been a large increase in studies analyzing the differences of microbiota in healthy subjects and in various disease states. Inflammatory bowel disease, liver disease, diabetes and obesity were the most explored so far, with a much larger spectrum still being researched. To promote bifidogenic effect, the usual recommendation for supplementation of inulin-type prebiotics is a daily dose of 2.5-10 g.

Based on all the studies that show the importance of prebiotic inulin as an adjunct in the treatment of several related comorbidities, the industry has been using inulin to improve the sensory quality and nutritional value of food products, among them, there were baked food, meat, sweets and dairy products. And scientists have been working jointly so that new products are developed using inulin as prebiotic. However, performed work to evaluate, through clinical trials, the effect in vivo prebiotic product developed, few found jobs, mostly with dairy products thus identifying the need for more development work to investigate, in practical tests, the effect of inulin in many products currently developed by the industry.

IMPORTANCE OF STUDYING MICROBIOTA

The microbiota terminology refers to a set of micro-organisms including bacteria, archaebacteria, viruses and some unicellular eukaryotes. In humans, it has been estimated that in 1014 CFU (colony forming unit) of microorganisms inhabiting various body parts such as the surface of the skin and gastrointestinal tract, respiratory and genitourinary [1].

A healthy microflora is defined as one that preserves and promotes the wellbeing and the lack of diseases, especially of the gastrointestinal tract. Under normal conditions, has significant influence on a number of biochemical reactions of the host such as dietary fiber processing to simple sugars, the processing of short chain fatty acids and other nutrients to be absorbed; Production of vitamin K, vitamin B12 and folic acid; participation in the metabolism and circulation of bile acids; transformation of potential carcinogens and activation of bioactive compounds [2].

Regarding microbial community that inhabits the gastrointestinal tract, this is a major microbial ecosystem in the presence of approximately 400 species of phylotypes. Each organ or portion of the system differs as to its anatomy and physiology which results in a chemical environment to the habitat of different bacteria. The concentration of microbial cells in the GI tract substantially increases the stomach to the colon [1, 3].

In the stomach there is increased production of acid, which is lethal to most micro-organisms, so only two genera are resistant to acids, *Lactobacillus* and *Streptococcus*, are prevalent. The secreted bile and pancreatic secretions in the upper duodenum and jejunum, but are toxic to most microorganisms, the number of bacteria progressively increases until 107 cells / ml in the distal ileum. The largest number of bacteria reside in the large intestine, this body which houses a complex and dense microbial community composed mainly of anaerobic species [4].

The colon is favorable environment for bacterial growth due to decreased transit available nutrients and a favorable pH (5.7 to 6.8). Generally, the bacteria which have almost exclusively sacarolitico metabolism, that is without proteolytic activity can be considered as potentially beneficial. This metabolic profile is observed in bacteria belonging to the genus *Lactobacillus* and *bifidobacteria* [5].

Most micro-organisms that inhabit the intestinal microbiota phyla *Firmicutes* belongs to (79.4%), *Bacteroidetes* (16.9%), *Actinobacteria* (2.5%) and *Proteobacteria* (1%), *Verrucomicrobia* (0.1%). *Archaea* and fungi are also present, but comprise less than 1% of the microorganisms [1, 6].

It is worth noting that the high interindividual variability in the intestinal microbiota composition makes it difficult to define a core microbiome (all members of a microbial ecosystem) that is shared by all human beings. Although the species that make up the intestinal flora of humans may vary in some circumstances, for example, acute diarrheal disease, antibiotic dietary interventions, the proportion of bacteria in the intestinal microbiota remains constant [3, 7].

The process of colonization of the gastrointestinal tract associated with the adherence of bacteria in the capacity of the intestinal mucosal adhesion receptors. The indigenous bacteria inhabiting the gastrointestinal tract continuously, with examples *bifidobacteria* and *lactobacilli,* whereas the non-native bacteria inhabit temporarily, since they do not have adequate adhesion to the gastrointestinal mucosa [8].

With the initial colonization process of the intestinal tract, it is emphasized that at birth, humans are essentially free of bacteria and the colonization process begins after birth and continues until the adult period. The initial microbiota is characterized by low diversity and consists primarily of facultative anaerobic bacteria belonging to the phyla Proteobacteria and *Actinobacteria*. This microflora becomes more diverse, and bacteria belonging to the phylum *Firmicutes* and *Bacteroidetes* become dominant. Thus, the initial colonization is important for the composition of the intestinal microbiota in adulthood [9, 10].

According to experimental data is possible to identify three main functions of the intestinal flora: 1) metabolism function as a result of biochemical activity of microorganisms, formation of short chain fatty acids, acetate, propionate and butyrate, production of vitamins K and B12 and favorable effects on the absorption of calcium and iron in the colon; 2) protection function, preventing invasion by infectious agents or growth of micro-organism with potential pathogen; and 3) trophic role on the proliferation and differentiation of the intestinal epithelium and the development of the immune system modulation [7].

An important metabolic function of the intestinal microbiota is fermentation that occurs in the colon. The main substrates involved in this process are endogenous, such as mucus, and dietary carbohydrates which are not digested in the upper gastrointestinal tract, and include resistant starch polysaccharides, oligosaccharides and sugar alcohols. Additionally, dietary components that stimulate fermentation favors the increase in bacterial mass, and thus the faecal mass, promoting an increase in stool volume. It is estimated that about 30 g of bacteria are produced for each 100 g of fermented carbohydrates [11].

In the fermentation process, the bacteria of the colon using a variety of carbohydrate hydrolyzing enzymes to produce hydrogen, methane, carbon dioxide, lactate and short chain fatty acids, particularly acetate, propionate and butyrate. These fatty acids play a crucial role in the physiology of the colon, since they are the main energy source for enterocytes and colonocytes, stimulate epithelial cell proliferation and visceral blood flow and increase the absorption of sodium and water by reducing the osmotic load of the carbohydrate [12].

Fermentation and production of fatty acid short-chain pathogenic bacteria inhibit growth by reducing the luminal and faecal pH. A low pH reduces peptide degradation and the formation of toxic compounds, such as ammonia, amines, and phenolic compounds, but also decreases the activity of undesirable bacterial enzymes [11].

Regarding the protection function, intestinal microbiota plays a role of direct protection through physical barrier formed by cellular components of the vascular endothelium and epithelial cells. Under normal conditions, anaerobic microorganisms resident in the colon grow forming a layer on the surface of the intestinal wall limiting the growth and colonization of potentially pathogenic microorganisms [13].

The trophic function, this is performed by competition, mainly by short-chain fatty acids, participating in the growth and differentiation of cell epithelium, playing important physiological actions in hepatic regulation of fats and sugars and serving as energy substrate for colonocytes [7].

Among the short-chain fatty acids that play a trophic role stands out butyrate. It is the preferred fuel for colonic epithelial cells and plays an important role in regulating cell proliferation and differentiation, but also this acid stimulates the immunogenicity of cancer cells. Butyrate exerts an antiproliferative activity on many cell types, and there are studies showing preventive effects of butyrate in the development of colon cancer and adenoma [14].

Additionally, the gut microbiota as an immune barrier function of human body. At birth the immune system is immature, and develops after colonization of the intestinal microbiota. This microbiota in turn stimulates the increase of the number of Peyer's patches, the release of immunoglobulins (especially secretory immunoglobulin A), recognition of dendritic cells and coordinates proinflammatory and anti-inflammatory signs [15].

Therefore, the intestinal microflora is beneficial for the person when there symbiosis with the host. Some bacteria that make up this microbiota are always beneficial, non-pathogenic and protective of the intestinal mucosa, such as *bifidobacteria* and *lactobacilli*, while others can play both pathogenic effects, the beneficial effects to the host, as the example bacteroides [8, 16].

However, some factors may influence this process of colonization, such as maternal microbiota, diet (breast milk or infant formula); type of delivery (normal or cesarean); gestational age (term or preterm). The composition of the intestinal microbiota is also influenced by clinical and surgical interventions and various environmental factors, including the use of antibiotics, lifestyle, diet and hygiene habits. In addition, a recent study showed a direct relationship between ABO blood system and the intestinal microbiota. In this study, the group had a higher concentration of *Bifidobacterium spp*., The AB group higher concentration of *Bacteroides fragilis* group B and higher concentration *of Lactobacillus spp*. [17-19].

So it becomes more important to maintain healthy gut microflora, with the consumption of foods with bioactive compounds in its composition (inulin, fruooligossacarídeos, galacto-oligosaccharides, among others) and / or supplements are considered safe by naturally contained in food and microbiota human intestinal [20, 21] with specific action: Prebiotics, Probiotics and Symbiotics. To do so, set up by Prebiotics selectively fermented ingredients that allow specific changes both in composition and / or activity of the gastrointestinal microbiota, conferring benefits on the health and welfare of the host [22]. Probiotics are live microorganisms which when administered in adequate amounts, confer health benefits to the host. Have the symbiotic are appropriate combinations of prebiotics and probiotics. A synbiotic product exerts both a prebiotic effect as probiotic [23].

Microbiota in Health and Disease, Interindividual Differences, Normal Weight and Obesity

Front of entire characterization of the microbiota, different compositions and functions, it is possible to say that the microbiota activity goes beyond of causing benefits and harms to the host, which can be termed as an important regulator of metabolic and immune status. In recent decades, there has been a large increase in studies analyzing the differences of microbiota in healthy subjects and in various disease states. Inflammatory bowel disease, liver disease, intestinal disorders, diabetes and obesity were the most explored so far, with a much larger spectrum still being researched. In healthy individuals, or even in individuals who have in common the same disease, intestinal bacterial communities are diverse and variable from person to person [24].

There are still an intraindividual difference that depends of various environmental, and especially nutritional stimuli [25, 26]. Two multicenter studies in Europe have demonstrated the relationship of the composition of the microbiota with metabolic and inflammatory markers. The low bacterial richness, lower diversity of the bacterial population, was associated with worse metabolic and inflammatory states. Le Chatelier et al. (2013), both investigating obese and normal weight populations, found in individuals with low bacterial richness greater visceral adiposity, insulin resistance, dyslipidemia and more pro-inflammatory phenotype. The difference in intestinal microbiota between obese and normal weight have also been investigated, where early studies indicated specific differences in species, but several surveys have brought contrary results, making this approach not fully consistent. The full composition of the microbiota seems to play a greater change than obesity or overweight by itself, but the intestinal microbiota with low richness is more common in obese than in normal weight weights individuals [25, 26].

One study assessed the microbiota of completely distinct populations, a community of human hunter-gatherers, the Hadza of Tanzania, and italian urban controls. The Hadza had bigger population and higher levels of bacterial richness than Italians controls. Even in comparison with two groups of samples from rural communities in Africa, The Hadza also showed more unique features, demonstrating that each environmental stimulus modulates the microbiota in a specific way. Their Hunter-gatherers lifestyle led to some unique intestinal microbiota characteristics like absence of *Bifidobacterium* and differences related to sex, probably reflecting the sexual division of labor [27].

Many studies have been performed in experimental models to characterize the microbiota, and so far, it seems that the taxonomy differences in humans may be less impactful compared with the pronounced effects observed in rodents. Experimental models shows significant phylum-level alterations in of obesity/normal weight, disease/health and nutritional factors such as fiber and simple sugars intake, hypercaloric and high fat diets [28].

EUBIOSE AND DYSBIOSIS

The intestinal microbiota, in its healthy state, prevents potentially pathogenic microorganisms to exert their effects and to proliferate, while bacteria with beneficial functions to the host, in addition to produce bactericidal enzymes preventing adhesion of certain species, compete physically for colonization sites along the digestive tract, [2]. A healthy microbiota, also called eubiose state can be defined by a high bacterial richness, maintenance of their symbiotic functions to the host and having the ability to resist changes under physiological stress [29].

On the other hand, the imbalance of intestinal microbiota is called dysbiosis, disorder increasingly considered in the diagnosis of various diseases characterized by colonic dysfunction due to a change of the intestinal microbiota, where there is a predominance of pathogenic bacteria over beneficial bacteria [30]. The state of dysbiosis, besides the imbalance between protective and harmful bacteria population, there is a lower bacterial richness and the presence of other pathobionts [2].

INTESTINAL MICROBIOTA ASSESSMENT TECHNIQUES

To characterize the microbiota and analyze the differences and changes that may occur, different assessment techniques have been used and rapid improvements has been accomplished in recent decades. Different techniques can phylogenetically Identify and/or quantify the components of the gut microbiota. The most widely used techniques today are based on the extraction of DNA and amplification of 16S ribosomal RNA genes. The 16S rRNA genes are highly conserved between bacterial species, but vary in a manner which allows species identification. In the characterization of bacterial populations, genetic information about their metabolic functions have been explored and may further identify the impact of this shifts of the microbiota for the host [31].

MICROBIOTA, INFLAMMATION AND LIPOPOLYSACCHARIDE

The intestine, the organ with the greatest presence of bacterial populations, has the largest mass of lymphoid tissue in the body, the gut associated lymphoid tissue (GALT). This tissue through epithelial intestinal cells and innate immune receptors, influence the state and the immune responses of the host. Innate immune receptors like Toll-like receptors (TLRs) and NOD-like receptors (RLNs) are presents in the GALT and are targets to different stimuli in the mucosa, allowing immune cells to be modulate and to produce different inflammatory signals, both locally and systemically. This balance is important in food tolerance and in bacterial tolerance itself, in which the mucosal and systemic immunity in balance can distinguish and respond appropriately to pathogens [29].

States of dysbiosis can break down this balance, where besides changes in bacterial populations, an increase of harmful bacteria (that may be followed by an increase in gram-negative bacteria), there are also other deleterious factors such as changes in intestinal permeability and barrier. The bacterial lipopolysaccharide (LPS) appears to be one of the most evident parameter which produces harmful effects through qualitative and quantitative changes in bacteria populations. The LPS, component of the cell wall of gram-negative, it is an endotoxin capable of induce an inflammatory response, this component is absorbed through the intestinal capillaries when the death of gram-negative bacteria occurs. In individuals with healthy microbiota, circulating LPS are in low concentrations, but in dysbiosis, there is an increased intestinal permeability caused by the reduction and modification of proteins junction membrane of epithelial cells, which may facilitate a greater flow of LPS to circulation, wherein high concentrations, cause a sustained pro-inflammatory stimuli [32].

The endotoxemia, elevated plasma concentrations of LPS, has been linked to inflammation with metabolic disorders such as insulin resistance and hepatic steatosis. The molecular mechanism that orchestrates these disorders is mainly signaling through CD14 and Toll-like receptor 4 (TLR4) into macrophages, monocyte cells and other innate immune complex, activating transcription of pro-inflammatory genes factors such as nuclear factor kappa-B and thereby promote an increase in pro-inflammatory cytokines [32, 33].

EFFECT OF DIET ON INTESTINAL MICROBIOTA

The main and most direct factor that influences the intestinal microbiota is food intake. Part of ingested food reaches the large intestine where the microbiota have a central role for the metabolism of dietary fibers, parts of plants that do not suffer the effects of digestive enzymes. Dietary fibers are key modulators of microbiota, but the complexity and uniqueness of the intestinal microbiota is a result of all intake and various other environmental factors. In animals, herbivores have with higher bacterial richness, longer intestinal transit and increased intestinal tracts. In humans, the intestinal microbiota may also have specific characteristics according to eating habits. The microbiota of a vegetarian has little capacity to metabolize carnitine, which is abundantly present in red meat, while the lower supply of amino acids in the vegetarians diet, makes their microbiota carry more genes for nitrogen assimilation [34, 35].

In humans, small changes in food intake are capable of influencing the population of the intestinal microbiota. The amount of energy and type of the most abundant nutrient in the diet produce this change. A trial with two groups of daily food intake of 2400 and 3400 kcal, showed significant changes in populations of bacterial genera and the ability to extract calories from food [36].

Another study showed that children in rural areas of Burkina Faso, who consumed higher amounts of plant polysaccharides had higher levels of Prevotella and Xylanibacter known to contain a set of bacterial genes for cellulose and xylan hydrolysis, as well as higher concentrations of SCFA than Italian children [37].

Each nutrient or food can promote specific changes in intestinal microbiota. The fiber intake affect bacterial populations directly by the increased supply of substrate to bacteria that can digest resistant starch and generate energy, which provides them a growth advantage in the gut. These changes in the composition of bacterial populations is also accompanied by an increase in concentrations SFCA and a decrease of the pH [38].

Changes may occur by the type of fiber ingested. In a double-blind, crossover trial, the intake of crackers enriched by different types of resistant starch (RS) showed results that RS2 and RS4 show functional differences in their effect on human fecal microbiota composition after 10 days of intervention, indicating that the chemical structure of RS may be of greater use for certain bacterial groups [39].

Some studies explored the effect of diets rich in certain macronutrients. Habitually high intakes of animal protein and fat leads to enterotypes rich in fecal Bacteroides, while diets rich in carbohydrates were associated with fecal enterotypes rich in Prevotella [40].

The hypo and hyperlipidemic diets have been used to elucidate various mechanisms and differences in the gut microbiota in various experimental models and also in humans. A diet rich in lipids exert strong selective pressure on the gut microbiota structure and function by stimulating the synthesis and enterohepatic circulation of primary bile acids because only bacteria that are able to tolerate the physiologic concentrations of bile acids in the gut survives [41].

The consumption of certain phytochemicals can also cause changes in the microbiota. Tzounis et al. (2013) reported that daily consumption of a high-flavanol cocoa drink (494 mg/day) for 4 weeks significantly increased *bifidobacterial* and *lactobacilli* and decreased clostridia histolyticum population counts compared to those consuming the low -cocoa flavonol drink (23 mg/day) [42].

Opon bacterial incubation, phenolic components and polyphenols exert positive effects which can inhibit pathogen growth and stimulate the growth of commensal bacteria. Grape pomace phenolic extract (1 mg/ml) [43] increases *Lactobacillus acidophilus* CECT 903 in growth media and culture liquid tea phenolics can reduce the proliferation of pathogens such as Clostridium spp. (C. perfringens and C. difficile) and gram-negative bacteria Bacteroides spp., where the same effect is reduced in commensal populations with beneficial effects [44].

In addition to the food intake, drugs can cause extreme shifts in the intestinal microbiota. Intake of antibiotic cause changes in bacterial populations that exert profound physiological and immunological responses in the host in which they still have little attention in the clinic. Besides the changes during antibiotic therapy, studies show that these changes are sustained for longer, Jernberg et al. (2007) demonstrated that a ignificant reduction in diversity of *Bacteroide*s persisted up to 2 years following 7 days of Clindamycin administration [45].

The use of antibiotics in diseases related to states of disbiosys is still subject of much study, they are effective in eradicating specific pathogens, but also reduce microbial diversity, reducing beneficial commensal bacteria and enabling opportunistic bacteria to colonize the newly hospitable niches in the gut ecosystem. In individuals with Crohn's disease (CD), the use of ciprofloxacin has been shown to modestly improve symptoms and remission rates of patients [46], but the same antibiotic exposure in childhood has been asssociated with development of inflammatory bowel diseases (IBD) in adulthood [47].

OBESITY

The difference in intestinal microbiota between obese and normal weight individuals have been extensively studied in the last decade, mainly due to the fact that in obesity, in general, there are changes in the food intake. Initially, the evidence showed an increase in the *Firmicutes-Bacteroidetes* gender relation in individuals with obesity, but recent studies have demonstrated that the difference between low and high bacterial richness phenotypes seem to have more important influences. Low bacteria richness correlates wealth with obesity, insulin resistance, dyslipidemia and more pronounced inflammatory state [26, 48].

Comorbidities associated with obesity can have both its genesis and its progress influenced by microbial communities. Thus, interventions that can modulate the intestinal microbiota can be used in trials in people with obesity to improve metabolic parameters and regulate the individual immune response. A systematic review and meta-analysis given the effect of prebiotics and synbiotics in a pooled analysis of clinical trials only with overweight and obesity. Interventions were able to improve lipid parameters such as decreased plasma concentrations of total cholesterol, LDL-C and triglycerides as well as improvements in glucose homeostasis, such as decreased plasma insulin concentrations [49].

The difference of the intestinal microbiota and the relation with body weight is also evidenced in children. Kalliomäki et al. (2008), in a prospective study that analyzed the microbiota of 49 children between 6 and 12 years of age, where the children that developed overweight at 7 years of age had higher concentrations of Staphylococcus aureus and lower levels of *Bifidobacteria* when were younger [50].

Diabetes Mellitus

Diabetes mellitus also has its peculiarities which were observed in relation to intestinal microbiota. The type 2 diabetes mellitus (T2DM) in which its physiological pillar it is insulin resistance (IR), has as main enhancer of condition the inclination the immune system to a pro-inflammatory state. Cell mechanism that involves the development of IR based on the attenuation of the transducing signal receptors for insulin as seen in adipocytes, myocytes, hepatocytes and beta cells. Changing one or more steps of the insulin signaling cascade may be caused by the increase of LPS and pro-inflammatory cytokines. The substrate receptor substrate 1 (IRS-1), responsible for transmitting signals to the intracellular environment, is phosphorylated on serine by enzymes which have their activity increased due to the increase of pro-inflammatory cytokines [51, 52].

Trials that used interventions to modulate the intestinal microbiota in individuals with T2DM showed metabolic improvements that were also positively correlated with changes in specific microbes (Bifidobacterium spp., *Lactobacillus* spp., and A. F. prausnitzii muciniphila) [53, 54]. Dewulf et al. (2013) showed that plasma concentrations of LPS, which are also increased in diabetic subjects, negatively correlated with phyla and species of short-chain fatty-acid producing bacteria such as *Firmicutes*, *Actinobacteria*, *Faecalibacterium prausnitzii* and *Bifidobacterium*. Still, the increase in Clostridium cluster IV was negatively correlated with fasting insulin and glucose, while a positive correlation between *Propionibacterium, Bacteroides intestinalis*, vulgates *Bacteroides* and fasting glucose was found [53].

Two other trials support that the intestinal microbiota influences and may have a related paper showing development of T2DM subjects whom were characterized by a reduction in *Roseburia species* and *Faecalibacterium prausnitzii,* which produce the SCFA butyrate [55, 56].

Diabetes mellitus type 1 (DM1), with a distinct pathophysiology of T2DM, which is an autoimmune disease leading to destruction of insulin secreting b-cells in the pancreas, has also shown a relationship between intestinal microbiota and its etiology [57]. Factors influencing the incidence of DM1 as breast feeding period, infant formula and their different types, early/late exposure to gluten and others foods, are also factors that directly modulate the intestinal microbiota of children. Despite the genetic development of the disease, the correlation between monozygotic twins is only 50%, making intriguing the research on mechanisms which make these other factors increases the incidence of disease. Children born with B-cell autoimmunity differ in their intestinal microbiota matched controls without autoimmunity [58].

Some authors have found changes in the concentrations of genera and species of Bacteroides, but not all results have found the same significant changes, however the populations may indicate a protective role of genres such as butyrate producers and Faecalibacterium Roseburia and mucin degraders like Prevotella and Akkermansia, which were found in higher concentrations in controls than in cases of T1DM [58, 59].

Non-Alcoholic Fat Liver Disease (NAFLD)

The liver and its important functions in digestion and metabolic control are intimately connected with the composition of the intestinal microbiota. The development of nonalcoholic fatty liver disease (NAFLD) has been related to changes in bacterial populations and to states of dysbiosis. Functions of microbiota like efficiency of energy harvest, healthy epithelial maintenance and flow of microorganisms and components as LPS in the portal circulation seems to be the most important factors in development of NAFLD. Activation of pattern-recognition receptors on hepatic cells by microbial-derived ligands results in the stimulation of innate immune and profibrotic pathways, such as the LPS triggering effect orchestred by TLR-4 / CD14 complex [60].

Clinical trials using modulators of intestinal microbiota have also been explored to improve several markers of the state of NAFLD, where a meta-analysis of the pooled data to check the efficacy of therapies showed that could significantly decreased alanine aminotransferase, aspartate transaminase, total cholesterol, high density lipoprotein, tumor necrosis factor-α and homeostasis model assessment of insulin resistance [61].

Inflammatory Bowel Diseases (IBD) and Irritable Bowel Syndrome (IBS)

The first diseases to have the intestinal microbiota explored for possible therapeutic treatments were the inflammatory bowel diseases (IBD), due to the fact that almost all diseases in this group have some form of colonic dysfunction that can reach extreme cases of dysbiosis [62].

Crohn's disease (CD) and ulcerative colitis (UC) are the most common IBD, both are chronic diseases mediated by immune factors and characterized by periods of remission and relapse during life. The colonization of different bacterial populations that regulate host suggests that immune responses to infection, autoimmunity, cancer, or other immunological conditions can be influenced by these microorganisms shifts [62]. Patients with IBD have quantitative and qualitative changes in intestinal microbiota composition characterized by low bacterial richness, density of Increased mucosal surface colonization and invasion in epithelial areas with active disease [63].

A study including microbial gene content from 231 subjects Identified enrichment in microbial pathways for oxidative stress tolerance, immune evasion, and host metabolite uptake, with corresponding depletions in SCFA biosynthesis [64]. It is unclear whether the changes to the microbiota described in these studies are a causal factor or merely a consequence of intestinal inflammation and IBD [62].

Some studies have found specific changes in the microbiota of patients with CD. Willing et al. (2010) [65] demonstrated the absence of certain bacterial species as Faecalibacterium and Roseburia, as well as an increase in other *Ruminococcus gnavus* and *Enterobacteriaceae spp.*. Sokol et al. (2008) [66] found a decrease in species that are potential anti-inflammatories such as *Faecalibacterium prausnitzii*.

Another very common disease involving the gastrointestinal tract, irritable bowel syndrome (IBS), is characterized by abdominal pain lasting more than 6 months and altered bowel habits without any organic cause, with a prevalence much higher than IBD, being about 10-20% worldwide [67].

IBS has been linked to three different specific conditions: mucosal dysfunction, low grade inflammation and changes in intestinal microbiota. Depending on the type of IBS, which can be diarrhea and/or constipation predominant, the microbiota of these patients exhibit specific changes [68].

For example, the genus *Lactobacillus* have been found increased or decreased in patients with IBS, in addition, another genus *Veillonella* has also been reported to increase its population in IBS. The microbiota modulation can be a good tool considering that the main symptoms in IBS as visceral hypersensitivity are directly related bacterial populations and their production of metabolites and gases [69, 70].

Airway Diseases

The respiratory tract also harbors a microbiota and the microbial composition of the airway microbiota is different between healthy lungs and those with chronic lung diseases, such as asthma, chronic obstructive pulmonary disease as well as cystic fibrosis [71]. The development of the intestinal microbiota of airline routes and share common factors that can modulate both from birth as the breastfeeding and early solid feeding [72].

Concerning intestinal microbiota, differences in the concentrations of the genus *Bifidobacterium* have been linked to asthma patients [73].

Renal Diseases

The kidney-gut axis has been increasingly investigated with chronic kidney disease (CKD), where two major physiopathological concepts that relate CKD with certain colonic disorders and altered intestinal permeability. The first builds on the production and accumulation of toxic metabolites derived from bacterial fermentation of protein and Increased substance with nitrogen in the GI tract, whereas the second concept is the translocation of endotoxins like LPS or bacteria from the very gut lumen into the bloodstream, due to damage of the intestinal epithelial barrier and quantitative and/or qualitative alterations of the intestinal microbiota associated with the uraemic milieu. These two circumstances are important to patients with CKD and can trigger chronic inflammation, increase cardiovascular risks and worsen uraemic toxicity [74].

The CKD promotes this changes due to various factors such as reduced intake of fiber owing to restrictions of potassium intake [75], the higher prevalence of constipation caused by dialysis, inactivity and pharmacotherapy, increased-protein substrate for bacterial proteolytic species due to uraemia impaired protein assimilition [76], changes in the intestinal pH due to increased blood ammonia concentration, and finally, medications like antibiotics, phosphate binders and antimetabolites [77].

MICROBIOTA AS A THERAPEUTIC TARGET

The event that orchestrates all the effects that prebiotics, probiotics and synbiotics may exert on the host it is microbiota modulation, wherein the benefits are associated with an increase of bacterial richness, followed by higher concentrations of beneficial bacteria and lower concentrations of harmful bacteria.

The supplementation of prebiotics, probiotics and synbiotics have been suggested as an useful tool for modulating states of dysbiosis in many cases, acting as an adjuvant therapy in restoring the homeostasis of certain metabolic pathways [26, 48, 78].

Despite the apparent stability of the intestinal microbiota in adults, there is a large intra and interindividual variability of intestinal microbiota lifelong [79]. This variability is dependent on several factors as seen previously, which can promote health or predispose and aggravate a number of diseases, such as obesity, diabetes and fatty liver disease [80]. Thus, there is a growing interest in intestinal microbiota, since it plays an important role both in health and in disease states. It has been well established that some bacterial genera exert positive health effects, such as *lactobacilli* and *bifidobacteria*, therefore, is reasonable to assume that the modulation of the intestinal microbiota is a useful strategy to promote growth and development of these bacteria [81-83]. Among the modulators of the intestinal microbiota there are the prebiotics, supplements considered safe because they are naturally present in food [84, 85]. Evidence shows that prebiotics exert bifidogenic effect in humans subjects, in other words, stimulate the growth of *bifidobacteria* in the gut [86, 87]. Regarding *lactobacilli*, studies have also shown the ability of these bacteria to use prebiotics as energetic substrate [88, 89].

PREBIOTICS: DEFINITION AND CHARACTERIZATION

A dietary prebiotic was defined two decades ago as a *"selectively fermented ingredient that results in specific changes in the composition and/or activity of the gastrointestinal microbiota, thus conferring benefits upon host health"* [22].

To be considered prebiotic, some requirements must be met [16]:

1) be neither hydrolyzed nor absorbed in the upper part of the gastrointestinal tract;
2) be a selective substrate for one or a limited number of beneficial bacteria commensal to the colon, which are stimulated to grow and/or are metabolically activated;
3) consequently, be able to alter the colonic flora in favor of a healthier composition;
4) induce luminal or systemic effects that are beneficial to the host health.

A variety of substances are considered prebiotics, but the most common and used as prebiotics are inulin, fructo-oligosaccharides (FOS) and oligofructose, all of them belonging to the chemical group of inulin-type fructans (ITF).

STRUCTURE AND DIETARY SOURCES OF INULIN-TYPE FRUCTANS

Inulin-type fructans are oligo- and polysaccharides of D-fructose joined by β(2–1) bonds with an α(1–2) linked D-glucose at the terminal end of the molecule [91]. The main difference between inulin, oligofructose and FOS is the degree of polymerization (DP), i.e., the number of individual monosaccharide units that make up the molecule. The commercial form of FOS is obtained by enzymatic addition of fructose residues to sucrose, with the DP varying between 2 and 4 fructose residues. The oligofructose is obtained from mixes produced by partial hydrolysis of inulin, consisting in 2 to 8 fructose residues with a terminal glucose residue or a chain of three to eight fructose residues. Inulin may contain between 2 and 60 fructose residues [91].

ITF are found naturally in several foods such as onion, Jerusalem artichoke, chicory, leek, garlic, artichoke, banana, rye, barley, dandelion, yacon and wheat (Table 1). However, these foods contain small amounts of ITF and the overall intake in a Western-type diet is also small. It is estimated that the daily intake of ITF is between 3 and 11 g/d in Europe, and between 1 and 4 g/d in North America [92]. The food industry has taken the initiative to add these active ingredients in food products most frequently consumed in order to reach levels where a prebiotic effect can occur (5-10 g/d) [88]. Inulin-type fructans are used as functional food ingredients in cereals, confectionery, biscuits, infant feeds, yogurts, table spreads, bread, sauces and drinks, as well as in dietary supplements [93].

Table 1. Quantity of inulin (%) in natural foods

Food	Edible part	Inulin (%)
Onion	Bulb	2-6
Jerusalem Artichoke	Tubercle	16-20
Chicory	Root	15-20
Leek	Bulb	3-10
Garlic	Bulb	9-16
Artichoke	Central leaves	3-10
Banana	Fruit	0.3-0.7
Rye	Cereal	0.5-1.0
Barley	Cereal	0.5-1.5
Dandelion	Leaves	12-15
Yacon	Root	3-19
Wheat	Cereal	1-4

Source: van Loo J et al. (1995) [92] and Moshfegh AJ et al. (1999) [94].

HEALTH EFFECTS RELATED TO THE INGESTION OF INULIN-TYPE FRUCTANS

Supported by the results of a large number of animal studies and human intervention trials, the interest in the benefits of inulin-type fructans has grown among various organizations and health professionals. In 2008, the Food and Agriculture Organization of the United Nations (FAO) convened a technical meeting [84] to start work on the evaluation of the functional and health properties of prebiotics. Likewise, in 2011, the World Gastroenterology Organisation (WGO) published guidelines [95] for the use of probiotics and prebiotics globally by gastroenterologists and other health professionals. In 2013, the International Life Sciences Institute (ILSI) published a monograph [96] that describe the concepts of probiotics and prebiotics for use in the human diet and explore the scientific basis for potential human health benefits. In 2014, a WGO handbook [97] provided an overview of the intestinal microbiota, its role in health and disease, probiotics and prebiotics. The FAO technical meeting recommended that prebiotic producers, medical professionals, and public health officers consider some form of system to monitor the health outcomes of long-term prebiotic administration. The WGO and ILSI indicate that there is evidence of efficacy for probiotics and prebiotics and suggest that these ingredients can be readily incorporated into a balanced diet.

Especifically for inulin-type fructans, evidence for health benefits include: reduction of plasma concentrations of pro-inflammatory markers, intestinal permeability and metabolic endotoxemia; improvement of lipid and glucose metabolism, mineral absorption, intestinal microbiota composition, gastrointestinal function and body weight management; increase of anorexigenic peptides, satiety and short-chain fatty acids production (Table 2).

Considering the scientific evidence presented, the benefits provided by consumption of inulin-type fructans makes the intestinal microbiota an important therapeutic target in management of various diseases and at the maintenance and promotion of health. The data demonstrate the importance of including inulin-type fructans in public health policies to discuss modifications in current clinical practice. However, there is still no consensus on the minimum daily dose to achieve beneficial effects on human health or the maximum tolerable dose without causing adverse effects, although some recommendations have been adopted.

Recommended Doses

To promote bifidogenic effect, the usual recommendation for supplementation of inulin-type prebiotics is a daily dose of 2.5-10 g. Bouhnik et al. [87] evaluated the ability of different doses of FOS to stimulate the growth of *bifidobacteria* in a 7-d study of 40 healthy human volunteers and reported that the optimal dose without significant side effects was 10 g/d. In this study, it was observed a linear dose-response relation at doses from 2.5 to 10 g/d, which suggests a dose-effect relationship.

Nevertheless, it has been questioned whether there is a dose-response relationship, since the composition of the intestinal microbiota of each individual (particularly *bifidobacteria*) is a major determinant of the effectiveness of the prebiotic, not only the dose [16, 128]. Studies have shown that small doses of ITF are capable of promoting bifidogenic effect [129, 130], but there is a great variability between individuals with respect to the composition of the intestinal microbiota, therefore, the response is also variable, which shows that other factors than the dose may influence individual response.

Table 2. Health benefits provided by supplementation of inulin-type fructans. Studies include only randomized clinical trials and meta-analyses

Outcomes	Health benefits
Immune function	Reduction of plasma concentrations of CRP [98], TNF-α [98, 99] and interleukin-6 [99].
Lipid metabolism	Reduction in free-fatty acids [101], in triglycerides [101, 102], total cholesterol [103], LDL-c [103] and HDL-c [102].
Glucose homeostasis	Reduction in fasting plasma glucose [99, 100, 103], glycated hemoglobin [98, 99], insulin [99, 101, 102-104] and HOMA-IR [99, 104], hepatic glucose production [105].
Intestinal permeability	Reduction in serum concentrations of zonulin [106], in lactulose-mannitol excretion ratio [106].
Metabolic endotoxemia	Reduction in plasma LPS [98, 99, 107].
Intestinal microbiota composition	Increase in fecal *Bifidobacteria* [107-117], *Lactobacilli* [108, 113, 114] and Faecalibacterium prausnitzii concentrations [107, 118]. Reduction in *Bacteroides/Prevotella* ratio [112, 113], *Bacteroides* spp. [34] *Propionibacterium* [32] and in *Clostridia* [116] and in fecal pH [110].
Gastrointestinal function	Softer stools [110], longer gastric emptying time [119], increase in stool frequency [107, 112], in the amount of feces [120] and in fecal wet weight [115]. Lower chance of developing diarrhea [118].
Mineral absorption	Increase in total absorbed calcium [121-125], magnesium [125], daily calcium accretion to the skeleton [123], bone formation [124], whole-body bone mineral content and density [123].
Weight management	Reduction in body weight [102, 104], waist circumference [105] and body mass index [104].
Satiety	Increase in satiety [104, 126]. Reduction in energy intake [102, 129] and in hunger [126].
Gastrointestinal peptides	Increase in plasma concentrations neurotensin [120], somatostatin [119], GLP-1 [100, 127], GLP-2 [106], PYY [102, 127]. Reduction in ghrelin [100, 102].
Short-chain fatty acids production	Increase in total SCFA [116], acetate [116], propionate [116] and butyrate [100, 116].

Abbreviations: CRP – C-Reactive Protein. GLP-1 - Glucagon-like peptide-1. GLP-2 – Glucagon-like peptide-2. HDL-c – High Density Lipoprotein Cholesterol. HOMA-IR – Homeostasis Model Assessment – Insulin Resistance. LDL-c – Low Density Lipoprotein Cholesterol. LPS – Lipopolysaccharide. PYY – Peptide YY. SCFA – Short-chain fatty acids. TNF-α – Tumor Necrosis Factor alpha.

INULIN

The inulin is the best-known and studied prebiotic due to its wide application in different industry sectors, in addition to being a renewable, inexpensive and abundant raw material. Inulin has a number of biotechnological applications, such as the production of bioethanol, fructose syrup, single-cell protein (substitute for protein-rich foods), single cell oil (transesterification of triacylglycerols from renewable biomass for the production of biodiesel), citric acid (used in the food, chemical cosmetics, beverage and pharmaceutical industries, as stabilizer for vegetable oils and fats or as complex forming and bleaching component in many washing detergents), obtainment of oligofructose (which has wide applications in various types of foods like confectionery, fruit preparations, milk desserts, yogurt and fresh cheese, baked goods, chocolate, ice cream and sauces), and other chemicals compounds [79].

Because of all the benefits provided by inulin, it becomes important to study the increasing role of this prebiotic, both in human health as well as in industrial applications.

The main function of fructans is the reserve carbohydrate of plants and are hydrolyzed when they require energy. There are other physiological functions that have been postulated by other researchers, but these have not been fully elucidated. Fructans increase the osmotic pressure of the plants cells, resulting in protection of the same against abiotic stress. A wide variety of carbohydrates are involved in this group and it is mainly the hexoses (glucose and fructose), disaccharides (sucrose and trehalose), sugar alcohols (e.g., mannitol and inositol), complex carbohydrates (eg raffinose and stachyose). The presence of fructans enables the tolerance to frost and as a result contributes to the stabilization of the membrane. The variation in the polymerization of grain can be induced by climatic conditions, thus fructans are evaluated for their biological role as a photosynthetic adaptation to low temperatures [131].

Fructans are a class of compounds which includes all plant polysaccharides and oligosaccharides, wherein the fructosyl-fructose linkages constitute the majority of glycosidic bonds; therefore, they are mainly polymers of fructose units. The fructans can have at least one fructosyl-glucose linkage, identical to those found in sucrose and, when present, is typically a round link chain of the polymer [132].

The oligosaccharide polysaccharides are characterized by a degree of polymerisation (DP), the type and sequence of their monosaccharide moieties.

Where the oligosaccharides to 10 monomeric units are contained in the chain which may be straight or branched. Since polysaccharides are hydrolysable carbohydrate formed by more than ten molecules of monosaccharides linked together by glycosidic linkages, forming long linear or branched chains. The hydrolysis of these compounds gives high amounts of monosaccharides, unlike the mono and disaccharides, these molecules are insoluble in water but has no effect on cell osmotic equilibrium. The GP is defined as the number of fructose units linked to a terminal glucose moiety [133, 134].

Among the non-digestible oligosaccharides are compounds containing glucose, xylose, galactose and fructose, or fructooligosaccharides (FOS). In most cases, the inulin molecule has a terminal glucose moiety (GFN) but can only have fructose molecules (Fm) and the terminal portion has no glucose. FOS are generally regarded as safe for human consumption. They are not hydrolyzed by digestive enzymes in the upper gastrointestinal tract and, due to its non-digestible, they reach the colon intact. The chain length, the presence of β-linkage and branching makes these compounds resistant to hydrolysis by digestive enzymes in the gastrointestinal tract and are therefore classified as natural fiber. Are more soluble than sucrose and contribute to the volume, improve umectância and provide crispness to low-fat products. Thus they play the same role of fat, but have the advantages of having lower caloric value, enrich the fiber content and other nutritional properties of food [92, 133-135].

Inulin is a polydisperse frutana with a degree of polymerization ranging from 2 to 60 fructose units, consisting of a mixture of higher oligomers and linear polymers of fructose. The β-D-fructofuranosyl units are held together by β- linkages of the type (2, 1) and have a glucose molecule at the initial linear portion of each chain of fructose, which is joined by a link type (α1- β-2) as the molecule of sucrose. Due polydispersity, molecular weight of inulin is between 3.500 and ± 5.500 [92, 135].

The inulin extracted from chicory is legally classified as a food ingredient (not as an additive) in Brazil and in all countries of the European Union as well as Switzerland and Norway. The authorities in Australia, Canada, Israel, Japan and New Zealand reached the same conclusion. In the United States, was confirmed GRAS status (Generally Recognized as Safe) to inulin and oligofructose [92].

Two of the main sources for industry are chicory (*C. intybus*) and Jerusalem artichoke (H. tuberosus). However, the main limitation of chicory as a source of inulin is the presence of 1-fructan exo-hydrolase, an inulinase degrading fructosyl-fructose linkages at low temperatures, reducing the quality of inulin in the final period of growth during and storage. The average chain length of inulin ranges from plants and in different growth conditions [134].

In some plants, for example, *C. intybus*, the degree of polymerization is determined by the type of cultivation, the plant life cycle and harvest time. The early harvest yields inulin with a higher degree of polymerisation compared to the late harvest, before the inulinases that long molecules can cleave short chain oligosaccharides [132].

Native inulin is extracted from a form of inulin from plant sources with hot water, hydrolysis minimizing any activity which may lead to their degradation. Commercial preparations containing inulin which are obtained from other forms than the native should not be considered as inulin product. The composition of the native GF n inulin molecules is accompanied by a small percentage of Fm molecules [134].

Various methods for the extraction of inulin from various sources have been documented. The process of extraction with hot water at 85 ° C is one of the most popular methods, followed by evaporation and spray drying. Another effective method for extraction of inulin, which is widely used in laboratory scale is precipitation from aqueous solution using ethanol. This method proved to be uneconomical and is not suitable for industrial production thereof [136].

Complex polysaccharides are difficult to be synthesized and handled, there are currently no commercial process to synthesize these compounds in an automated manner, which poses a great limitation to the detailed study of their biological functions. Furthermore, there is no method available for determining structural biologists to carbohydrate DP <10 in comparison with other biological molecules such as nucleic acids can be synthesized in vitro using the polymerase chain reaction (PCR) [136].

Despite these limitations, there are several methods that can be used for the synthesis of fructose and fructo-oligosaccharides with a DP <10 from a simple low cost raw materials. Fructose can be obtained by acid hydrolysis of inulin at high temperatures, although the monosaccharide product can be easily degraded to low pH and the process may give rise to staining of the inulin hydrolyzate and formation of undesired by-product, which decrease the yield and require extensive processing [136].

The hydrolysis of inulin using enzymes is usually carried out at temperature of 60 ° C and neutral pH. The temperature is especially critical since it prevents the microbial contamination of the final product decreases the viscosity, improving transfer rates and allows the use of higher concentrations of inulin due to increased solubility.

These conditions are crucial to employ inulinolíticas highly thermostable enzyme, eg a thermostable extracellular exoinulinase *Aspergillus fumigatus* and a stable exoinulinase heat from *Streptomyces sp.* For the production IOS, the main advantage of using a higher reaction temperature lies in the increase of the concentration of the substrate that improves the performance of IOS [134].

INULIN AS PREBIOTIC

Bifidogenic activity is described as the most effective with short chain FOS in which fructosyl units (n = 2-8) have a connection β- (2, 1). Inulin fermentation properties of multiplication of *Bifidobacterium longum, B. infantis and B. angulatumi*, increased their levels of 20-71% and decreased the levels of clostridia 3 to 0.3% [137]. The bifidobacterium proliferation is associated with beneficial effects of inulin on health such as dietary fiber according to strengthen defense mechanisms to assist in the loss of excess weight, diabetes control, improve immune function of the gut, improved absorption mineral, beneficially in lipid metabolism, have anticancer effects, among other health claims still need more research to validate it, especially with humans [138].

The dietary fiber term was coined in 1953, but the health benefits of foods rich in fiber have been used for a long time. 430 BC Hippocrates described in wheat laxative effects "large" compared to refined wheat. In the 1920s, JH Kellogg has published extensively on the attributes of wheat bran, alleging that the increase in stool weight, promoting laxative effect, and preventing diseases. Dietary fiber has been researched over the years 1930 and after relatively forgotten until the 1970s [139].

Denis Burkitt was widely credited for again spread the idea that dietary fiber protects against development of Western diseases, including diabetes, cardiovascular disease, colon cancer, and obesity. Since then continue the investigation into the definition, types of fibers, and determination of consumption for health benefits. Cohort studies clearly show that high fiber intake is associated with decreased cardiovascular disease. The data on obesity, diabetes and cancer are more variable. Dietary fiber is listed on the Nutrition Facts table on the food label, and the recommendations are for every 100 g of food, 2.5 g fiber (to be considered good source of fiber) and 5.0 g (to be considered excellent source of fiber) [139].

Future studies should distinguish between health benefits derived exclusively from the consumption of soluble fiber and associated with the growth and activity of the intestinal microbiota. Future intervention studies exploring the effect of prebiotics in the diet on immune function must be performed in healthy individuals who are subsequently exposed to an immune challenge [139].

INULIN AS IMMUNOMODULATION

Inulin is known for its ability to stimulate the growth of beneficial bacteria while suppressing the growth of potentially harmful microorganisms, such beneficial relationship is given by the development of a low pH environment, increased affinity for the cell wall adhesion, increased availability nutrients, production of deleterious metabolites for pathogenic microorganisms (for example, short chain fatty acids) and immunomodulatory (mucin production and bacteriocin) [30, 32, 33].

Intestinal flora, innate immune function and metabolism are closely linked with onset of pathological processes at the molecular level (subclinical inflammation, immune cell activation, increased oxidative stress and endoplasmic reticulum, altered production of vascular adhesion molecules and glycation products advanced) that contribute to the eventual development of metabolic disorders such as hyperlipidemia, atherosclerosis, insulin resistance and weight gain [33, 32, 35, 36, 39].

Colonizing bacteria fermentation of prebiotics and their products may play a key role in immunomodulation, both by increased host resistance to infection and inappropriate immune responses of chronic inflammatory conditions or for maintaining the integrity of the gastrointestinal barrier. Prebiotics reduce the invasion of pathogenic bacteria and their products (including LPS) for the circulation, preventing activation of downstream immune cells [33].

Currently there is insufficient evidence to recommend the inulin as a modulator of immune function to improve health. There are very few clinical trials available, and reported contradictory results. Although individual results are found in significant increases in measures of antioxidant capacity and reductions in circulating concentrations of LPS in the small intestine after prebiotics interventions, further studies are needed to verify these results. Inulin has antioxidant properties regardless of the change of the growth of bacteria in the intestine and is capable of decreasing the number of reactive oxygen species, which can help reduce the lipid peroxidation in the stomach [32-34].

Association with Body Weight

In addition, proposed that metabolic endotoxemia (ie, increased blood levels of lipopolysaccharide (LPS)) triggers the low-grade inflammation that occurs with overeating, obesity and diabetes and thus contributes to these metabolic disorders [32-36].

In humans, obesity is characterized by dysbiosis. Initial studies demonstrated that changes in the proportion and / or levels of predominant bacterial phyla (a decrease in the Bacteroides and / or an increase in Firmicutes) occur in obese individuals, and can be reversed by diet and weight loss, but these changes are still to be a matter of debate. Changes also occur in the genus level, such as least amount of *Bifidobacteria* in obese compared with normal-weight and diabetes when compared to non-diabetic individuals [49, 102].

The SCFA acetate, propionate and butyrate are produced as by-products of prebiotic bacterial fermentation. Besides representing an energy source to the host, these SCFA perform a variety of beneficial functions including the maintenance of human health and modulation of intestinal metabolic and immunological processes. SCFA are the only known ligands for two G-protein coupled receptor, GPR41 and GPR43, which is expressed on a variety of gastrointestinal cells and stimulates secretion of hormones involved in the regulation of energy intake and expenditure [102, 126].

The connection for the AGCC GPR41 peptide YY increases production and GLP-1 hormones that reduce appetite, slowing gastric emptying and increasing insulin sensitivity. SCFA also promote differentiation of intestinal L cells, contributing to the increase of endogenous GLP-1 production [102, 126].

When analyzed four clinical trials in a meta-analysis, individually, two found a higher concentration of GLP-1 after supplementation with prebiotics in healthy subjects and overweight. However, the concentration of GLP-1 was not significant after the meta-analysis. Each of the trials reported significant reductions in ghrelin concentrations and increased GLP-2 concentrations in subjects who used inulin [102, 126].

Most of the studies investigating the effect of inulin supplementation on regulation of intestinal peptide (peptide YY), and secretion of the incretin

(GLP-1) reported a significant increase in production of these molecules after 2 weeks, but the changes were not significant when analyzed in meta-analysis. The role played by certain bacteria and prebiotics in the gut hormone kinetics require further investigation, as dietary fibers (not probiotics) have also been reported to be associated with increased SCFA, peptide YY and GLP-1 [102, 126].

In five trials investigating the effect of the prebiotic supplementation on self-assessments on satiety three showed improvements in subjective measures of satiety in healthy subjects consuming probiotics when compared to the three controls and found a significant reduction in total energy consumption during the intervention with prebiotic compared with the group that was supplemented with placebo. However, the reduction in energy consumption lost statistical significance after meta-analysis. In addition, the consensus found among the three tests mentioned above, the consumption of a prebiotic for at least 2 weeks increasing satiety signals in healthy adults were supported on results based on self-reports [132].

Increases in the production of hydrogen expired in the intervention group, indicating higher intestinal bacterial fermentation were detected by one of the trials investigating the effects of prebiotic supplementation in feelings of satiety. However, changes in hunger and satiety reported by the subjects in these studies may have occurred regardless of any changes in the gut microbial fermentation because soluble fiber and inulin, it is able to modify intestinal transit [132].

Contradictory results have been reported in clinical trials to examine the effect of the prebiotic action in body weight, significant reductions in body weight after prebiotic supplementation compared to placebo were observed in two of the five studies, while no change in body weight was observed in three trials. Have been unlikely to be observed in trials with two weeks because of the short duration of their activities, long-term studies (12-17 weeks) were the most likely to observe reductions in body weight than shorter essays [132].

As for reducing energy consumption, two high-quality clinical trials provide consistent evidence favoring the consumption of prebiotic in at least 2-week diet to reduce total energy consumption in normal weight and overweight and in women with diabetes type 2. a supplementation period with 12 weeks was required before low power consumption of the participants resulted in significant weight reduction. However, the reduction in energy consumption was not statistically significant when these two trials were part of a meta-analysis, with three more trials of high quality [53, 132].

Where body composition was assessed, the weight lost was predominantly fat mass rather than lean body mass. Long-term trials are now needed to determine whether the consumption of prebiotics in the diet is a safe and effective therapeutic option for reducing weight and body fat in the long term or if the individual physiological adaptations, would eventually compensate for this energy imbalance to minimize loss weight [49, 102, 132].

More studies are needed to correlate these findings with changes in the growth and function of specific intestinal bacteria. At present there is insufficient evidence for inulin supplementation recommendation to reduce power consumption and body weight by increasing gastric peptide YY and the secretion of GLP-1, improving insulin sensitivity, lowering lipid levels and modulating immune function, and interesting to conduct prospective studies of long-term, for the analysis of the role of physiological repercussion of this supplementation [102].

Insulin Resistance, Diabetes Control and Glucose Intolerance

The composition of the intestinal microbiota and the possible effects on human health have been the focus of scientific interest, as they are being developed and updated metagenomic techniques that allow the identification and characterization of intestinal bacteria that can not be grown. Furthermore, the discovery of differences in the composition of the intestinal microbiota between individuals and groups in obese and normal subjects with and without type 2 diabetes, highlighted the potential role of the microflora of the colon and their fermentation products in the pathogenesis of metabolic health host and disease [49, 53-57, 99, 102].

Although the number and diversity of bacterial species within the gastrointestinal tract of an individual remains relatively constant over the life, it is possible to stimulate the proliferation of specific micro-organisms known to have beneficial health effects through manipulating the dietary host [49, 53-57, 99, 102].

It has been shown that a daily intake of inulin significantly reduced basal hepatic glucose production in healthy humans with no effect on the insulin metabolism. After a meta-analysis of four studies that measured the effect of inulin supplementation on concentrations of postprandial glucose, two reported significant reductions in blood glucose in obese and normal weight women. Also, after the meta statistically significant effect was observed, confirming that the prebiotic intake results in reducing postprandial glucose concentrations are positive [49, 53-57, 99, 102].

However, a study investigating fasting glucose, fasting insulin concentrations and insulin resistance (HOMA-IR) reported conflicting results. Significant reductions in glycated hemoglobin (HbA1c) in healthy subjects after only five weeks of supplementation with inulin and women with type 2 diabetes after eight weeks were found, while for supplemental obese women with duration of three months there was no change in HbA1c [100, 104].

The provision of the inulin type supplements can restore the growth of beneficial bacterial species and improve intestinal barrier integrity. Studies in animals have shown a causal relationship between the consumption of a diet high in fat and increased levels of intestinal bacteria containing LPS, or LPS circulating concentrations and the development of obesity and insulin resistance. In the configuration of a high-fat diet, LPS is capable of translocating from the gut to the host circulation, resulting in "Metabolic endotoxemia." LPS stimulates the excessive production of reactive oxygen species and pro-inflammatory cytokines by macrophages, resulting in subclinical systemic inflammation, weight gain and insulin resistance development [99, 100].

Human studies have found that indvíduos with type 2 diabetes have endotoxemia plasma levels twice as high as the levels observed in non-diabetic controls. Metabolic endotoxemia is also positively correlated with the total energy consumption and fasting insulin in non-diabetic population. Inulin and its fermentation products were analyzed by reducing gastrointestinal permeability by a variety of mechanisms [49, 53-57, 99, 102].

The results of studies investigating the fasting glucose and fasting insulin concentrations and insulin resistance (HOMA-IR) after supplementation with prebiotics were contradictory. Prebiotic long-term intervention studies in persons with pre-diabetes or metabolic syndrome are now required to determine whether probiotics confer some protection against type 2 diabetes future development in high-risk individuals [49, 53-57, 99, 102, 104].

Prebiotics and Non-Alcoholic Fat Liver Disease

NAFLD is characterized by an asymptomatic fatty infiltration of the liver and inflammation which can eventually lead to fibrosis, cirrhosis, portal hypertension, hepatocellular carcinoma and liver failure. The obesity, dyslipidemia, insulin resistance and diabetes have been reported to be associated with the development of NAFLD. Higher plasma endotoxemia, overproduction of inflammatory cytokines and oxidative excessive stress inside the liver cells are also thought to contribute to the pathogenesis of NAFLD [61, 62].

The use of probiotic dietary supplements to restore optimal microbial balance in the gastrointestinal tract of individuals with NAFLD can help reduce the TAG accumulation in the liver, promote inflammation and attenuate hepatic secretion of lipoproteins such as VLDL. translocation of gram-negative bacteria from the gut into the circulation has been reported to be associated with increased severity of cirrhosis. By maintaining gut barrier function and reduces bacterial translocation, prebiotics may be effective in treating liver disease complications [61, 62].

Studies exploring the effects of dietary intake of prebiotic in metabolic parameters in patients with NASH are scarce. In this review, a small study involving seven adult male was included, which found a significant reduction in aspartate aminotransferase concentrations after prebiotic supplementation. This conclusion was supported by a larger randomized controlled trial (n 66), who administered a symbiotic (fructo-oligosaccharides Bifidobacterium longum þ) to participants for 24 weeks. In addition to a significant reduction in aspartate aminotransferase concentrations, researchers found a decrease in circulating levels of cytokines (TNF-a) and inflammation (C-reactive protein), reduced serum concentrations of LDL-cholesterol and Benefits metabolic prebiotics in the diet British Journal of Nutrition 11 endotoxin, an improvement in insulin sensitivity (HOMA-IR) and reduced hepatic steatosis (determined by liver biopsy). More research is needed in this potentially very promising area of study [61, 62, 132, 134].

ROLE IN MINERAL ABSORPTION

Research conducted in animal models have shown that non-digestible oligosaccharides (NDO) such as inulin, oligofructose and transgalactooligosaccharide (TOS) stimulate the absorption of minerals, especially calcium and magnesium. This paper FOS was demonstrated by the accumulation of bone mineral content and also the prevention of osteoporosis in ovariectomized rats by the addition of 5% FOS led to a significant increase in bone mineral content. Increased bone mineralization is possibly as a result of improving mineral passive and active transport across the intestinal epithelium mediated by an increase in the intestinal flora metabolites and a decrease in pH. Moreover, inulin has been shown to increase Fe uptake in rats with iron deficiency anemia [122-125].

Role in Lipid Metabolism

In addition to its beneficial effect on the gastrointestinal tract, studies in animal models have shown that oligofructose and inulin modified hepatic lipid metabolism. By feeding rats with inulin or oligofructose, it was demonstrated that these carbohydrate diets markedly reduced serum triacylglycerols (TAG) and phospholipid concentrations

Outcomes associated with cardiovascular and hepatic health [45-49].

There was sufficient evidence to support the prebiotic supplementation lowers LDL cholesterol or total concentration in healthy, obese or dyslipidemic, with most studies finding no change in the concentrations of these lipids after the intervention. Of the eleven trials investigating the effect of probiotic supplementation on the concentrations of circulating TAG five reported significant reductions in healthy individuals with overweight or hypercholesterolemia compared with controls. However, the remaining six tests failed to detect changes in TAG concentrations were also conducted in healthy individuals with overweight or hypercholesterolemia [45-49].

Inulin: Modulation and Prospects in New Products

Based on all the studies that show the importance of prebiotic inulin as an adjunct in the treatment of several related comorbidities, the industry has been using inulin to improve the sensory quality and nutritional value of food products, and scientists have been working jointly so that new products are developed using inulin as prebiotic.

In a systematic review of products with inulin found 256 articles that spoke of inulin and product development, assessing the improvement of sensory and nutritional quality of the product with inulin compared to the standard product. Among between food products were, baked goods, meats, sweets and dairy products. The largest group was presented the dairy products, among which were observed work also evaluated the symbiotic function of the product, ie to meet their prebiotic inulin function multiplied bacteria also added that produto [140].

However, performed work to evaluate, through clinical trials, the effect in vivo prebiotic product developed, few found jobs, mostly with dairy products [141-143], thus identifying the need for more development work to investigate, in practical tests, the effect of inulin in many products currently developed by the industry.

References

[1] Harris K. et al. Is the gut microbiota anew factor contributing to obesity and its metabolic disorders? *Journal of Obesity*, 2012.

[2] DiBaise JK, Zhang H, Crowell MD, Krajmalnik-Brown R, Decker GA, Rittmann BE, et al. Gut microbiota and its possible relationship with obesity. *May Clin Proc*. 2008; 83: 460-469.

[3] Blaut M, Klaus S. Intestinal Microbiota and Obesity. *Handbook of experimental pharmacology*. 2012; p. 251-273.

[4] Aron-Wisnewsky J, Doré, J, Clement K. The importance of the gut microbiota after bariatric surgery. *Nat Rev Gastroenterology e Hepatology***.** 2012; 10: 590-598.

[5] Cummings JH, Macfarlane GT. The control and consequences of bacterial fermentation in the human colon. *J Appl Bacteriol*. 1991; 70: 443-459.

[6] Tap J. Towards the human intestinal microbiota phylogenetic core. *Environ Microbiol*. 2009; 11: 2574-2584.

[7] Guarner F, Malagelada JR. Gut flora in health and disease. *Lancet*. 2003; 9356: 512-529.

[8] Brandt KG, Sampaio MMSC, Miuki CJ. Importância da microflora intestinal. *Pediatria*. 2006; 28: 117-127.

[9] Carlisle EM, Morowitz MJ. Pediatric surgery and the human microbiome. *J Pediatr Surg***.** 2010; 46: 577-584.

[10]Backhed F. ProGramming of Host Metabolism by the Gut Microbiota. *Ann Nutr Metab*. 2011; 58: 44-52.

[11]Slavin J. Fiber and prebiotics: mechanisms and health benefits. *Nutrients*. 2013; 5: 1417-1435.

[12]Almeida LB, Marinho CB, Souza CS, Cheib VBP. Disbiose intestinal. *Rev Bras Nutr Clin*. 2009; 24:58-65.

[13]Bourlioux P, Koletzko B, Guarner F, Braesco V. The intestine and its microflora are partners for the protection of the host: report on the Danone

Symposium "The Intelligent Intestine," held in Paris, June 14, 2002. *Am J Clin Nutr.* 2003; 78: 675-683.

[14]Wong JNW. Colonic health: fermentation and short chain fatty acids. *J Clin Gastroenterol.* 2006; 40: 235-43.

[15]Tappenden KA, Deutsch AS. The physiological relevance of the intestinal microbiota--contributions to human health. *J American Coll Nutr.* 2007; 26: 679S-683S.

[16]Gibson GR, Roberfroid MB. Dietary modulation of the human colonic microbiota: introducing the concept of prebiotics. *J Nutr.* 1995**;** 25: 1401-1412.

[17]Marchesi JR. Human distal gut microbiome. *Environ Microbiol.* 2011**;** 13: 3088-3102.

[18]Makivuokko H, Lahtinen SJ, Wacklin P, Tuovinen E, Tenkanen H, Nikkila J, Bjorklund M, Aranko K, Ouwerhand AC, Matto J. Association between the ABO blood group and the human intestinal microbiota composition. *Microbiol.* 2012; 6: 12-94.

[19]Sommer F, Backhed F. The gut microbiota--masters of host development and physiology. *Nature Rev Microbiol.* 2013; 11: 227-238.

[20]Pineiro M, Asp N, Reid G, Macfarlane S, Morelli L, Brunser O, Tuohy K. FAO Technical meeting on prebiotics. *J Clin Gastroenterol.* 2008; 42: 156S-159S.

[21]Sanders ME, Akkermans LMA, Haller D, Hammerman C, Heimbach JT, Hormannsperger G, Huys G. Safety assessment of probiotics for human use. *Gut Microbes***.** 2010; 1: 164-185.

[22]Gibson GR, Probert HM, Loo JV, Rastall RA, Roberfroid MB. Dietary modulation of the human colonic microbiota: updating the concept of prebiotics. *Nutr Res Rev.* 2004; 17(2):259-75.

[23]23 WGO (WORLD GASTROENTEROLOGY ORGANISATION). Probiotics and Prebiotics. 2011. Disponível em: <http://www.worldgastroenterology.org/assets/export/userfiles/Probiotics_FINAL_20111128.pdf>. Acesso em: 09 abr. 2014.

[24]Human Microbiome Project, C. Structure, Function and diversity of the healthy human microbiome. *Nat.* 2012; 486: 207-214.

[25]Cotillard A, Sean PK, Ling CK, Edi P, Nicolas P, Le Chatelier E, et al. Dietary intervention impact on gut microbialgene richness. *Nat.* 2013; 500. Avaiable online doi:10.1038/nature12480.

[26]Le Chatelier E, Nielsen T, Junjie Q, Edi P, Falk H, Gwen F, et al. Richness of human gut microbiome correlates with metabolic markers. *Nat.* 2013; 500: 541-550.

[27]27Schnorr SL, Candela M, Rampelli S, Centanni M, Consolandi C, Basaglia G, et al. Gut microbiome of the Hadza hunter-gatherers. *Nat. Comm.* 2014; 5: 3654.

[28]Wu GD, Chen J, Hoffmann C, et al. Linking long-term dietary patterns with gut microbial enterotypes. *Scien.* 2011; 334: 105-108.

[29]James MK, Alexander CVR, Elaine H, Ara D, Jeremy KN. The human gut microbiome: Implications for future health care. *Curr Gastr Rep.* 2008; 10: 396-403.

[30]Yee KC, Mehrbod E, Deanna LG. Clinical Consequences of Diet-Induced Dysbiosis. *Ann Nutr Metab.* 2013; 63(suppl 2):28–40.

[31]Marianne HF, Paul WO, Eamonn MMQ. Techniques used to characterize the gut microbiota: a guide for the clinician. Nat Rev Gastroenterol Hepatol 2012; 9: 312–322.

[32]Cani PD, Amar J, Iglesias MA, Poggi M, Knauf C, Bastelica D, et al. Metabolic endotoxemia initiates obesity and insulin resistance. *Diab.* 2007; 6: 1761–1772.

[33]Chow JC, Young DW, Golenbock DT, Christ WJ, Gusovsky F. Toll-like receptor-4 mediates lipopolysaccharide-induced signal transduction. *The Journ of Biologic Chemestr.*1999; 274: 689-692.

[34]Muegge BD, Kuczynski J, Knights D, Clemente JC, González A, Fontana L, et al. Diet drives convergence in gut microbiome functions across mammalian phylogeny and Within humans. *Scienc.* 2011; 332: 970-974.

[35]Robert AK, Zeneng W, Bruce SL, Jennifer AB, Elin O, Brendan TS, et al. Metabolism of intestinal microbiota L-carnitine, a nutrient in red meat, promotes atherosclerosis. *Nat Med.* 2013; 19: 576-585.

[36]Jumpertz LR, Le DS, Turnbaugh PJ, Trinidad C. Bogardus C, Gordon, JI, et al. Energy-balance studies reveal associations between gut microbes, caloric load, and nutrient, Energy-balance studies reveal associations between gut microbes, caloric load, and nutrient. *Am J Clin Nutr.* 2011; 94(1): 58-65.

[37]Carlotta F, Duccio C, Monica P, Matteo R, Jean BP, Sebastien M, et al. Impact of diet in shaping gut microbiota revealed by a comparative study in children from Europe and rural Africa. *Proc Natl Acad Sci.* 2010; 107: 14691-14696.

[38]Diane FB, Terri B, Suzanne H, Jay-Lin J, James H, Li L, et al. Resistant Starch: Promise for Improving Human Health. *Adv Nutr.* 2013; 4: 587-601.

[39]Inés M, Jaehyoung K, Patrick RD, Vicki LS, Jens W. Resistant starches types 2 and 4 have differential effects on the composition of the fecal microbiota in human subjects. *PLoS ONE*. 2010; 5: e15046.

[40]Gary DW, Jun C, Christian H, Kyle B, Ying-Yu C, Sue AK, et al. Linking long-term dietary patterns with gut microbial enterotypes. *Scienc*. 2011; 334: 105-108.

[41]Ridlon J, Kang DJ, Hylemon P. Bile salt biotransformations by human intestinal bacteria. *J. Lipid Res*. 2006, 47, 241-259.

[42]Tzounis X, Rodriguez-Mateos A, Vulevic J, Gibson GR, Kwik-Uribe C, Spencer JP. Prebiotic evaluation of cocoa-derived flavanols in healthy humans by using a randomized, controlled, doubleblind, crossover intervention study. *Am J Clin Nutr*. 2011; 93: 62-72.

[43]Hervert-Hernández D, Painted C, Rotger R, Goñi I. Stimulatory role of grape pomace polyphenols on Lactobacillus acidophilus growth. *Int. J. Food Microbiol*. 2009; 136: 119-122.

[44]Lee HC, Jenner AM, Low CS, Lee YK. Effect of tea Their phenolics and aromatic metabolites on bacterial intestinal fecal microbiota. *Res. Microbiol*. 2006; 157: 876-884.

[45]Jernberg C, Lofmark S, Edlund C, Jansson JK. Long-term ecological Impacts of antibiotic administration on the human intestinal microbiota. *ISME J*. 2007; 1: 56-66.

[46]Arnold GL, Beaves MR, Pryjdun VO, Mook WJ. Preliminary study of ciprofloxacin in active Crohn's disease. *Inflamm Bowel Dis*. 2002; 8: 10-15.

[47]Kronman MP, Zaoutis TE, Haynes K, Feng R, Coffin SE; Antibiotic exposure and IBD among children development: the population based cohort study. *Pediatrics*. 2012; 130: E803-e794.

[48]Cani PD, Geurts L, Matamoros S, Plovier H, Duparc T. Glucose metabolism: Focus on gut microbiota, the endocannabinoid system and beyond. *Diabet & Metab*. 2014; 40: 246-257.

[49]Beserra BTS, Fernandes R, Rosario VA, Mocellin MC, Kuntz MGF, Trindade EBSM. A systematic review and meta-analysis of the prebiotics and synbiotics effects on glycaemia, insulin concentrations and lipid parameters in adult patients with overweight or obesity. *Clin Nutr*. In press 2014.

[50]Kalliomäki M, Collado MC, Salminen S, Isolauri E. Differences in Early fecal microbiota composition in children may predict overweight. *Am J Clin Nutr*. 2008; 87: 534-538.

[51]Kahn SE, Hull RL, Utzschneider KM. Mechanisms linking obesity to insulin resistance and type 2 diabetes. *Natur.* 2006; 444: 840–846.

[52]Osborn O, Olefsky JM. The cellular and signaling networks linking the immunesystem and metabolism in disease. *Nat Med.* 2012; 18:363–374.

[53]Dewulf EM, Cani PD, Claus SP, Fuentes S, Puylaert PG, Neyrinck AM, et al. Insight into the prebiotic concept: lessons from an exploratory, double blind intervention study with inulin-type fructans in obese women. *Gut.* 2013; 62(8): 1112-21.

[54]Vulevic J, Juric A, Tzortzis G, Gibson GR. A mixture of trans-galactooligosaccharides angiograms markers of metabolic syndrome and modulates the faecal microbiota and immune function of adults overweight. *J Nutr.* 2013; 143 (3): 324-31

[55]Creely SJ, McTernan PG, Kusminski CM, Fisher M, Da Silva NF, Khanolkar M, et al. Lipopolysaccharide activates an innate immune system response in human adipose tissue in obesity and type 2 diabetes. *Ame Journ of Phys Endocr and Metab.* 2007; 292 (3): e740-7.

[56]Qin J, Li Y, Cai Z, Li S, Zhu J, Zhang F, et al. A metagenome-wide association study of gut microbiota in type 2 diabetes. *Nat.* 2012; 490: 55-60.

[57]Fredrik HK, Valentina T,Intawat N, Göran B, Carl JB, BjörnF, et al. Gut metagenome in European women with normal, impaired and diabetic glucose control. *Nat.* 2013; 498: 99-103.

[58]Goffau MC, Luopajarvi K, Knip M, Ilonen J, Ruohtula TT, Orivuori L, et al. Fecal microbiota composition differs between children with b-cell autoimmunity and those without. *Diab.* 2013; 62: 1238-1244

[59]Murri, M., Leiva, I., Gomez-Zumaquero, JM, Tinahones, FJ, Cardona, F., Soriguer, F, et al. Gut microbiota in children with Differs from type 1 diabetes que in healthy children: a case-control study. *BMC Med.* 2013; 21: 11:46.

[60]Wajahat ZM. The Gordian Knot of dysbiosis, obesity and NAFLD. *Nat Rev Gastroenterol Hepatol.* 2013; 10: 637–644.

[61]Yan-Yan M, Lin L, Chao-Hui Y, Zhe S, Li-Hua C, You-Ming L. Effects of probiotics on nonalcoholic fatty liver disease: A meta-analysis. *World J Gastroenterol.* 2013; 28: 6911-6918.

[62]Curtis H, Aleksandar DK, Ramnik JX. Inflammatory bowel disease as a model for translating the microbiome. *Immun.* 2014; 40: 843-854.

[63]Zhanju L, Anthony TC, Yingzi C. Microbiota regulation of inflammatory bowel disease and colorectal cancer. *Sem Cancer Biol.* 2013; 23p: 543–552.

[64]Morgan XC, Tickle TL, Sokol H, Gevers D, Devaney K, Ward DV, et al. Dysfunction of the intestinal microbiome in inflammatory bowel disease and treatment. *Genome Biol.* 2012; 13: R79.

[65]Willing BP, Dicksved J, Halfvarson J, Andersson AF, Lucio M, Zheng Z,et al. A pyrosequencing study in twins shows that gastrointestinal microbial profiles vary with inflammatory bowel disease phenotypes. *Gastroenterol.* 2010; 139: 1844–1854.

[66]Sokol H, Pigneur B, Watterlot L, Lakhdari O, Bermudez-Humaran LG, Grata-doux JJ, et al. Faecalibacterium prausnitzii is an anti-inflammatory commensal bacterium identified by gut microbiota analysis of Crohn disease patients. *Proc Natl Acad Sci.* 2008; 105: 16731–6.

[67]Longstreth GF, Thompson WG, Chey WD, Houghton LA, Mearin F, Spiller RC. Functional bowel disorders. *Gastroenter.* 2006; 130: 1480-91.

[68]Codling C, O'Mahony L, Shanahan F, Quigley MS, Marchesi JR. Molecular analysis of fecal and mucosal bacterial communities in irritable bowel syndrome. *Dig Dis Sci.* 2010; 55: 392-7.

[69]Malinen E, T Rinttila, Kajander K, Matto J Kassinen A Krogius L, et al. Analysis of the fecal microbiota of irritable bowel syndrome and healthy controls pacientes with real-time PCR. *Am J Gastroenterol.* 2005; 100: 373-82.

[70]Tana C, Umesaki Y, Imaoka A, Handa T, Kanazawa M, Fukudo S. Altered profiles of intestinal microbiota and organic acids may be the origin of Symptoms in irritable bowel syndrome. *Neurogastroenterol Motil.* 2010; 22: 512-9

[71]Eva SG, Benjamin JM. Microbiota abnormalities in inflammatory airway diseases — Potential for therapy. *Pharm & Ther.* 2014; 141 :32–39.

[72]Madan JC, Koestler DC, Stanton BA, Davidson L, Moulton LA, Housman, ML, et al. Serial analysis of the gut and respiratory microbiome in cystic fibrosis in infancy: interaction between intestinal and respiratory tracts and impact of nutritional exposures. *Bio.* 2012; 3: e00251-12.

[73]Ouwehand AC, Isolauri E, He F, Hashimoto H, Benno Y., Salminen, S. Differences in Bifidobacteriumflora composition in allergic and healthy infants. *J Allergy Clin Immunol.* 2001; 108: 144–145.

[74]Alice S, Giuseppe R, Irene B, Aderville C, Santo M, Enrico F. Alterations of intestinal barrier and microbiota in chronic kidney disease. *Nephrol Dial Transplant.* 2014 0: 1–10.

[75]Krishnamurthy VM, Wei G, Baird BC, Murtaugh M, Chonchol MB, Raphael KL, et al. High dietary fiber intake is associated with decreased

inflammation and all-cause mortality in patients with chronic kidney disease. *Kidney Int.* 2012; 81: 300-306.

[76]Bammens B, Verbeke K, Vanrenterghem Y, Evenepoel P. Evidence for impaired assimilation of protein in chronic renal failure. *Kidney Int.* 2003; 64: 2196-2203.

[77]Wong J, Piceno YM, Desantis TZ, Pahl M, Andersen GL, Vaziri ND. Expansion of urease and uricase containing, indole and p-cresol-forming and contraction of short chain intestinal fatty acid-producing bacteria in ESRD. *Am J Nephrol.* 2014; 39: 230-7.

[78]Guarner F, Khan AG, Garisch J, Eliakim R, Gangl A, Thomson A, et al. World Gastroenterology Organisation Global Guidelines: probiotics and prebiotics October 2011. *J Clin Gastroenterol.* 2012; 46: 468-81.

[79]Cheng J, Palva AM, de Vos WM, Satokari R. Contribution of the intestinal microbiota to human health: from birth to 100 years of age. *Curr Top Microbiol Immunol.* 2013; 358: 323-246.

[80]Fukuda S, Ohno H. Gut microbiome and metabolic diseases. *Semin Immunopathol.* 2014; 36(1): 103-114.

[81]Bernardeau M, Vernoux JP, Henri-Dubernet S, Guéguen M. Safety assessment of dairy microorganisms: the Lactobacillus genus. *Int J Food Microbiol.* 2008; 126(3):278-285.

[82]Russell DA, Ross RP, Fitzgerald GF, Stanton C. Metabolic activities and probiotic potential of bifidobacteria. Int J Food Microbiol. 2011; 149(1): 88-105.

[83]Williams NT. Probiotics. *Am J Health Syst Pharm.* 2010; 15: 67(6):449-458.

[84]Pineiro M, Asp NG, Reid G, Macfarlane S, Morelli L, Brunser O, et al. FAO Technical meeting on prebiotics. *J Clin Gastroenterol.* 2008; 42 Suppl 3 Pt 2:S156-159.

[85]Sanders ME, Akkermans LM, Haller D, Hammerman C, Heimbach J, Hörmannsperger G, et al. Safety assessment of probiotics for human use. *Gut Microbes.* 2010; 1(3): 164-185.

[86]Falony G, Lazidou K, Verschaeren A, Weckx S, Maes D, De Vuyst L. In vitro kinetic analysis of fermentation of prebiotic inulin-type fructans by Bifidobacterium species reveals four different phenotypes. *Appl Environ Microbiol.* 2009 Jan;75(2): 454-461.

[87]Bouhnik Y, Vahedi K, Achour L, Attar A, Salfati J, Pochart P, et al. Short-chain fructo-oligosaccharide administration dose-dependently increases fecal bifidobacteria in healthy humans. *J Nutr.* 1999; 129(1): 113-116.

[88]Bañuelos O, Fernández L, Corral JM, Valdivieso-Ugarte M, Adrio JL, Velasco J, et al. Metabolism of prebiotic products containing beta(2-1) fructan mixtures by two Lactobacillus strains. *Anaerobe.* 2008; 14(3): 184-189.

[89]Endo H, Tamura K, Fukasawa T, Kanegae M, Koga J. Comparison of fructooligosaccharide utilization by Lactobacillus and Bacteroides species. *Biosci Biotechnol Biochem.* 2012; 76(1): 176-179.

[90]Kolida S, Gibson GR. Prebiotic capacity of inulin-type fructans. *J Nutr.* 2007; 137(11 Suppl): 2503S-2506S.

[91]Roberfroid MB. Introducing inulin-type fructans. *Br J Nutr.* 2005; 93 Suppl 1: S13-25.

[92]van Loo J, Coussement P, de Leenheer L, Hoebregs H, Smits G. On the presence of inulin and oligofructose as natural ingredients in the western diet. *Crit Rev Food Sci Nutr.* 1995; 35(6): 525-552.

[93]Moshfegh AJ, Friday JE, Goldman JP, Ahuja JK. Presence of Inulin and Oligofructose in Diets of Americans. *J Nutr.* 1999; 129(7 Suppl): 1407S-11S.

[94]Guarner F, Khan AG, Garisch J, Eliakim R, Gangl A, Thomson A, et al. World Gastroenterology Organisation Global Guidelines: probiotics and prebiotics October 2011. *J Clin Gastroenterol.* 2012; 46(6): 468-481.

[95]Binns N. Probiotics, Prebiotics and the Gut Microbiota. *ILSI Europe Concise Monograph Series* 2013:1-32.

[96]World Gastroenterology Organisation. *WGO Handbook on Gut Microbes.* 2014 May:1-66.

[97]Dehghan P, Gargari BP, Jafar-Abadi MA, Aliasgharzadeh A. Inulin controls inflammation and metabolic endotoxemia in women with type 2 diabetes mellitus: a randomized-controlled clinical trial. *Int J Food Sci Nutr.* 2014; 65(1): 117-123.

[98]Dehghan P, Pourghassem Gargari B, Asghari Jafar-abadi M. Oligofructose-enriched inulin improves some inflammatory markers and metabolic endotoxemia in women with type 2 diabetes mellitus: a randomized controlled clinical trial. *Nutrition.* 2014; 30(4): 418-423.

[99]Tarini J, Wolever TM. The fermentable fibre inulin increases postprandial serum short-chain fatty acids and reduces free-fatty acids and ghrelin in healthy subjects. *Appl Physiol Nutr Metab.* 2010; 35(1): 9-16.

[100] Jackson KG, Taylor GR, Clohessy AM, Williams CM. The effect of the daily intake of inulin on fasting lipid, insulin and glucose concentrations in middle-aged men and women. *Br J Nutr.* 1999; 82(1): 23-30.

[101] Brighenti F. Dietary fructans and serum triacylglycerols: a meta-analysis of randomized controlled trials. *J Nutr.* 2007; 137(11 Suppl): 2552S-2556S.

[102] Parnell JA, Reimer RA. Weight loss during oligofructose supplementation is associated with decreased ghrelin and increased peptide YY in overweight and obese adults. *Am J Clin Nutr.* 2009; 89(6): 1751-1759.

[103] Daubioul CA, Horsmans Y, Lambert P, Danse E, Delzenne NM. Effects of oligofructose on glucose and lipid metabolism in patients with nonalcoholic steatohepatitis: results of a pilot study. *Eur J Clin Nutr.* 2005; 59(5): 723-726.

[104] Genta S, Cabrera W, Habib N, Pons J, Carillo IM, Grau A, et al. Yacon syrup: beneficial effects on obesity and insulin resistance in humans. *Clin Nutr.* 2009; 28(2): 182-187.

[105] Luo J, Rizkalla SW, Alamowitch C, Boussairi A, Blayo A, Barry JL, et al. Chronic consumption of short-chain fructooligosaccharides by healthy subjects decreased basal hepatic glucose production but had no effect on insulin-stimulated glucose metabolism. Am J *Clin Nutr.* 1996; 63(6): 939-945.

[106] Russo F, Linsalata M, Clemente C, Chiloiro M, Orlando A, Marconi E, et al. Inulin-enriched pasta improves intestinal permeability and modifies the circulating levels of zonulin and glucagon-like peptide 2 in healthy young volunteers. *Nutr Res.* 2012; 32(12): 940-946.

[107] Dewulf EM, Cani PD, Claus SP, Fuentes S, Puylaert PG, Neyrinck AM, et al. Insight into the prebiotic concept: lessons from an exploratory, double blind intervention study with inulin-type fructans in obese women. *Gut.* 2013; 62(8): 1112-1121.

[108] García-Peris P, Velasco C, Lozano MA, Moreno Y, Paron L, de la Cuerda C, et al. Effect of a mixture of inulin and fructo-oligosaccharide on Lactobacillus and Bifidobacterium intestinal microbiota of patients receiving radiotherapy: a randomised, double-blind, placebo-controlled trial. *Nutr Hosp.* 2012; 27(6): 1908-1915.

[109] Closa-Monasterolo R, Gispert-Llaurado M, Luque V, Ferre N, Rubio-Torrents C, Zaragoza-Jordana M, et al. Safety and efficacy of inulin and oligofructose supplementation in infant formula: results from a randomized clinical trial. *Clin Nutr.* 2013; 32(6): 918-27.

[110] Petry N, Egli I, Chassard C, Lacroix C, Hurrell R. Inulin modifies the bifidobacteria population, fecal lactate concentration, and fecal pH but

does not influence iron absorption in women with low iron status. *Am J Clin Nutr.* 2012; 96(2): 325-331.

[111] Marteau P, Jacobs H, Cazaubiel M, Signoret C, Prevel JM, Housez B. Effects of chicory inulin in constipated elderly people: a double-blind controlled trial. *Int J Food Sci Nutr.* 2011; 62(2): 164-170.

[112] Kleessen B, Schwarz S, Boehm A, Fuhrmann H, Richter A, Henle T, et al. Jerusalem artichoke and chicory inulin in bakery products affect faecal microbiota of healthy volunteers. *Br J Nutr.* 2007; 98(3): 540-549.

[113] Costabile A, Kolida S, Klinder A, Gietl E, Bäuerlein M, Frohberg C, et al. A double-blind, placebo-controlled, cross-over study to establish the bifidogenic effect of a very-long-chain inulin extracted from globe artichoke (Cynara scolymus) in healthy human subjects. *Br J Nutr.* 2010; 104(7): 1007-1017.

[114] Ramnani P, Gaudier E, Bingham M, van Bruggen P, Tuohy KM, Gibson GR. Prebiotic effect of fruit and vegetable shots containing Jerusalem artichoke inulin: a human intervention study. *Br J Nutr.* 2010; 104(2): 233-240.

[115] Ten Bruggencate SJ, Bovee-Oudenhoven IM, Lettink-Wissink ML, Katan MB, van der Meer R. Dietary fructooligosaccharides affect intestinal barrier function in healthy men. *J Nutr.* 2006; 136(1): 70-74.

[116] Whelan K, Judd PA, Preedy VR, Simmering R, Jann A, Taylor MA. Fructooligosaccharides and fiber partially prevent the alterations in fecal microbiota and short-chain fatty acid concentrations caused by standard enteral formula in healthy humans. *J Nutr.* 2005; 135(8): 1896-1902.

[117] Lewis S, Burmeister S, Brazier J. Effect of the prebiotic oligofructose on relapse of Clostridium difficile-associated diarrhea: a randomized, controlled study. *Clin Gastroenterol Hepatol.* 2005; 3(5): 442-448.

[118] Ramirez-Farias C, Slezak K, Fuller Z, Duncan A, Holtrop G, Louis P. Effect of inulin on the human gut microbiota: stimulation of Bifidobacterium adolescentis and Faecalibacterium prausnitzii. *Br J Nutr.* 2009; 101(4): 541-550.

[119] Russo F, Clemente C, Linsalata M, Chiloiro M, Orlando A, Marconi E, et al. Effects of a diet with inulin-enriched pasta on gut peptides and gastric emptying rates in healthy young volunteers. *Eur J Nutr.* 2011; 50(4): 271-277.

[120] Kim SH, Lee da H, Meyer D. Supplementation of baby formula with native inulin has a prebiotic effect in formula-fed babies. *Asia Pac J Clin Nutr.* 2007; 16(1): 172-177.

[121] Abrams SA, Griffin IJ, Hawthorne KM. Young adolescents who respond to an inulin-type fructan substantially increase total absorbed calcium and daily calcium accretion to the skeleton. *J Nutr.* 2007; 137(11 Suppl): 2524S-2526S.

[122] Abrams SA, Griffin IJ, Hawthorne KM, Liang L, Gunn SK, Darlington G, et al. A combination of prebiotic short- and long-chain inulin-type fructans enhances calcium absorption and bone mineralization in young adolescents. *Am J Clin Nutr.* 2005; 82(2): 471-6.

[123] Holloway L, Moynihan S, Abrams SA, Kent K, Hsu AR, Friedlander AL. Effects of oligofructose-enriched inulin on intestinal absorption of calcium and magnesium and bone turnover markers in postmenopausal women. *Br J Nutr.* 2007; 97(2): 365-372.

[124] van den Heuvel EG, Muys T, van Dokkum W, Schaafsma G. Oligofructose stimulates calcium absorption in adolescents. *Am J Clin Nutr.* 1999; 69(3): 544-548.

[125] Cani PD, Joly E, Horsmans Y, Delzenne NM. Oligofructose promotes satiety in healthy human: a pilot study. *Eur J Clin Nutr.* 2006; 60(5): 567-572.

[126] Verhoef SP, Meyer D, Westerterp KR. Effects of oligofructose on appetite profile, glucagon-like peptide 1 and peptide YY3-36 concentrations and energy intake. *Br J Nutr.* 2011; 106(11): 1757-1762.

[127] Roberfroid M. Prebiotics: the concept revisited. *J Nutr.* 2007; 137(3 Suppl 2): 830S-7S.

[128] Rao VA. The prebiotic properties of oligofructose at low intake levels. *Nutr Res.* 2001; 21(6): 843-848.

[129] Bouhnik Y, Raskine L, Simoneau G, Paineau D, Bornet F. The capacity of short-chain fructo-oligosaccharides to stimulate faecal bifidobacteria: a dose-response relationship study in healthy humans. *Nutr J.* 2006; 28: 5-8.

[130] Chi ZM, Zhang T, Cao TS, Liu XY, Cui W, Zhao CH. Biotechnological potential of inulin for bioprocesses. *Bioresour Technol.* 2011; 102(6): 4295-4303.

[131] Apollinaris AC¸ Damasceno BPGL, Beltrão NEM, Pessoa A, Converti A, Silva JA. Inulin-type fructans: A review on different aspects of biochemical and pharmaceutical technology. *Carboh Polym.* 2014; 101: 368–378.

[132] Kelly, G. Inulin-Type Prebiotics – A Review: Part 1. *Altern Med Rev.* 2008; 13: 315-329.

[133] 133Lindhorst TK. *Essential of carbohydrate chemistry and biochemistry.* 3rd Ed. Wiley, Weinheim, Germany.

[134] Mutanda T, Mokoena MP, Olaniran AO, Wilhelmi BS, Whiteley CG. Microbial enzymatic production and applications of short-chain fructooligosaccharides and inulooligosaccharides: recent advances and current perspectives. *J Ind Microbiol Biotechnol*. 2014; 41: 893–906.

[135] SAAD, S. M. I. Probióticos e prebióticos: o estado da arte. *Rev Bras Ciên Farm*. 2006; 42: 1-16.

[136] Lingyun W, Jianhua W, Xiaodong Z, Da T, Yalin Y, Chenggang C, Tianhua F, Fan Z. Studies on the extracting technical conditions of inulin from Jerusalem artichoke tubers. J Food Eng. 2007; 79: 1087–1093.

[137] Waitzberg DL, Pereira CCA, Logullo L, Manzoni JT, Almeida D, Silva MLT, Torrinhas RSMM. Microbiota benefits after inulin and partially hydrolized guar gum supplementation – a randomized clinical trial in constipated women. *Nutr Hosp*. 2012; 27: 123-129.

[138] Hollister EB, Gao C, Versalovic J. Compositional and Functional Features of the Gastrointestinal Microbiome and Their Effects on Human Health. *Gastroenterol*. 2014; 146: 1449–1458.

[139] Slavin J. Fiber and Prebiotics: Mechanisms and Health Benefits. *Nutrients*. 2013; 5: 1417-1435.

[140] Kuntz MGF, Fiates GMR, Teixeira E. Characteristics of prebiotic food products containing inulin. *British F J*. 2013; 115: 235-251.

[141] Closa-Monasterolo R, Gispert-Llaurado M, Luque V, Ferre, Rubio-Torrents C, Zaragoza-Jordana M, Escribano J. Safety and efficacy of inulin and oligofructose supplementation in infant formula: Results from a randomized clinical trial. *Clin Nutr*. 2013; 32: 918-927.

[142] Abrams SA, Griffin IJ, Hawthorne KM, Liang L, Gunn SK, Darlington G, Kenneth Ellis KJ. A combination of prebiotic short- and long-chain inulin-type fructans enhances calcium absorption and bone mineralization in young adolescents. *Am J Clin Nutr*. 2005; 82: 471–476.

[143] Román JL, Martínez ABG, Luque A, Pons JAM, Vargas AA, Iglesias JR, Hernández M, Villegas JA. Efecto de la ingesta de un preparado lácteo con fibra dietética sobre El estreñimiento crónico primario idiopático. *Nutr Hosp*. 2008; 23(1): 12-19.

INDEX

A

accounting, 34
acid, viii, 104, 106, 107, 114, 123, 125, 140, 143
acidic, 21
acidity, 31
active transport, 132
adaptation, 123
adaptations, 130
adenoma, 107
adhesion, 15, 54, 105, 109, 127
adipose tissue, 138
adolescents, 144, 145
adulthood, 105, 113
adults, 118, 129, 138, 142
adverse effects, 121
Africa, 108, 136
age, 113, 140
alanine aminotransferase, 115
aldehydes, 31
amines, 106
amino acids, viii, 111
ammonia, 106, 117
amplitude, 7, 11
amylopectin retrogradation, vii, 2, 14, 35, 46
anaerobic bacteria, 105
anatomy, 104
anemia, 133
ANOVA, 10, 14
antibiotic, 105, 112, 113, 137
anti-inflammatories, 116
antioxidant, 3, 48, 127
appetite, 128, 144
aqueous suspension, 39
Asia, 143
aspartate, 115, 132
assessment, 109, 115, 135, 140
assessment techniques, 109
assets, 135
assimilation, 111, 140
asthma, 117
asymptomatic, 131
atherosclerosis, 127, 136
authorities, 124
autoimmune disease, 114
autoimmunity, 115, 116, 138

B

bacteria, 104, 105, 106, 107, 109, 110, 111, 112, 113, 114, 117, 118, 119, 127, 129, 130, 131, 132, 134, 137, 140
bacterial fermentation, 117, 128, 129, 134
bacterium, 139
Belgium, 5

beneficial effect, 107, 112, 121, 126, 133, 142
benefits, 3, 107, 108, 118, 120, 121, 122, 123, 126, 127, 134, 145
beverages, 52, 53
bile acids, 104, 112
biochemistry, vii, 144
biodiesel, 123
biomass, 123
biopolymer, 50
biopolymers, 48
biopsy, 132
biosynthesis, 116
biotechnological applications, 123
bleaching, 123
blends, vii, 1, 4, 5, 12, 19, 21, 23, 24, 36, 38, 46, 48, 49
blood, 106, 107, 117, 128, 130, 135
blood flow, 106
blood group, 135
bloodstream, 117
body composition, 130
body fat, 130
body mass index, 122
body weight, 113, 121, 122, 129, 130
bonds, 119, 123
bone, 122, 132, 144, 145
bone form, 122
bone mineral content, 122, 132
bowel, viii, 103, 108, 116, 138, 139
branching, 124
Brazil, 103, 124
breast feeding, 114
breast milk, 107
breastfeeding, 117
Burkina Faso, 111
by-products, 128

C

Ca^{2+}, 17
calcium, 105, 122, 132, 144, 145
cancer, 107, 116, 126
cancer cells, 107
carbohydrate, 31, 106, 123, 124, 125, 133, 144
carbohydrates, 106, 112, 123
carbon dioxide, 106
cardiovascular disease, 126
cardiovascular risk, 117
carotene, 22
carotenoids, 21
causal relationship, 131
CCA, 145
cellulose, 111
challenges, 54
cheese, 123
chemical, 38, 50, 104, 112, 119, 123
chemicals, 123
Chicago, 10
childhood, 113
children, 111, 113, 114, 136, 137, 138
Chile, 31
chlorophyll, 21
chloroplast, 21
cholesterol, 132, 133
chronic diseases, 116
chronic kidney disease (CKD), 117, 139, 140
chronic obstructive pulmonary disease, 117
chronic renal failure, 140
circulation, 104, 110, 112, 115, 127, 131, 132
cirrhosis, 131, 132
clinical trials, ix, 104, 113, 122, 127, 128, 129, 134
cocoa, 112, 137
colon, 3, 104, 105, 106, 107, 119, 124, 126, 130, 134
colon cancer, 107, 126
colonization, viii, 103, 105, 106, 107, 109, 116
color, 48, 54
colorectal cancer, 138
commercial, 3, 22, 41, 54, 119, 125
communities, 108, 139
community, 108
competition, 106

competitiveness, 2
complex carbohydrates, 123
complexity, 111
compliance, 15
complications, 132
composition, 3, 7, 11, 31, 36, 38, 48, 105, 107, 108, 111, 112, 115, 116, 117, 118, 119, 121, 122, 125, 130, 135, 137, 138, 139
compounds, 3, 31, 53, 104, 106, 107, 123, 124, 125
Concise, 141
configuration, 131
consensus, 121, 129
constipation, 116, 117
consumers, 2, 3, 29, 41, 54
consumption, 7, 107, 112, 121, 124, 126, 127, 129, 130, 131, 142
contamination, 125
controlled trials, 142
convergence, 136
cooking, 7
cooling, 15, 19
coronary heart disease, 3
correlation, 34, 41, 44, 114
correlations, 31, 41, 44
cosmetics, 123
cost, 125
crop, 52
CRP, 122
crystalline, 19, 39, 51
crystallisation, viii, 52
crystallites, 15, 39, 46
crystals, 15, 18, 38, 39
cultivation, 125
culture, 112
cystic fibrosis, 117, 139
cytokines, 111, 114, 131, 132

D

dairy products, ix, 29, 30, 104, 133, 134
data set, 10
defects, 31
defense mechanisms, 126
deficiency, 133
deformation, 11, 52
degradation, 106, 125
dendritic cell, 107
Denmark, 7, 10
destruction, 114
detectable, 23
detection, 34
detergents, 123
diabetes, viii, 103, 108, 118, 126, 128, 129, 131, 132
dialysis, 117
diarrhea, 116, 122, 143
diet, viii, 103, 107, 111, 112, 119, 120, 127, 128, 129, 130, 131, 132, 136, 141, 143
dietary fiber, 104, 111, 126, 129, 139
dietary intake, 132
digestibility, 3
digestion, 115
digestive enzymes, 111, 124
discriminant analysis, vii, 2, 10, 34
discrimination, 4, 33, 37, 46
disease-free, viii, 103
diseases, 104, 109, 113, 115, 118, 121, 126, 139, 140
disorder, 109
dispersion, 4
diversity, 105, 108, 112, 113, 130, 135
division of labor, 109
DNA, 109
dose-response relationship, 121, 144
drugs, 112
dry matter, 4
drying, 125
duodenum, 104
dyslipidemia, 108, 113, 132

E

ecosystem, 104, 105, 113
emulsions, 15, 21, 36, 50, 51, 52, 54
endothelium, 106
endotoxemia, 110, 121, 122, 128, 131, 132, 136, 141

endotoxins, 117
energy, 106, 111, 115, 122, 123, 128, 129, 130, 144
energy consumption, 129
entanglements, 11
environment, 104, 105, 114, 127
environmental factors, 107, 111
environments, 51
enzyme, 126
enzymes, 106, 109, 114, 124, 125
epithelial cells, 106, 110
epithelium, viii, 103, 106, 133
equilibrium, 124
ESRD, 140
ethanol, 125
ethyl alcohol, viii
etiology, 114
Europe, 108, 119, 136, 141
European Union, 124
evaporation, 125
evidence, 113, 121, 127, 129, 130, 133
EVOO, vii, 1, 3, 4, 5, 6, 10, 11, 12, 13, 14, 15, 16, 17, 18, 19, 20, 21, 22, 23, 24, 25, 26, 27, 28, 29, 31, 32, 33, 34, 35, 36, 37, 39, 40, 41, 42, 43, 44, 46
exclusion, 18
excretion, 122
experimental condition, 9, 33
exposure, 113, 114, 137
extra virgin olive oil, vii, 1, 3, 17, 20, 22, 25, 26, 27, 28, 32, 33, 34, 35, 37, 38, 39, 43, 46, 49, 53
extraction, 21, 109, 125

F

fasting, 114, 122, 131, 141
fasting glucose, 114, 131
fat, vii, 3, 5, 12, 18, 21, 24, 29, 30, 36, 38, 40, 41, 43, 44, 46, 47, 51, 52, 53, 54, 112, 124, 130, 131
fatty acids, 104, 105, 106, 121, 122, 127, 135, 141
FDA, 3
feces, 122
feelings, 129
fermentation, 106, 126, 127, 129, 130, 131, 135, 140
fiber, 109, 111, 112, 117, 124, 126, 127, 129, 143
fiber content, 124
fibrosis, 131
filler particles, 15, 46, 50, 51
fillers, 15
filtration, viii
financial support, 47
first dimension, viii, 2, 42
flavonol, 112
flavour, 3, 30, 31, 38, 41, 47
flocculation, 35
flora, 3, 119, 127, 134
fluid, 42
folic acid, 104
food, vii, ix, 2, 3, 4, 5, 10, 12, 38, 40, 41, 48, 50, 52, 104, 107, 110, 111, 112, 113, 118, 119, 123, 124, 126, 133, 145
Food and Drug Administration, 48
food industry, 3, 119
food intake, 111, 112, 113
food products, ix, 2, 41, 104, 119, 133, 145
formation, 13, 15, 17, 18, 19, 29, 105, 106, 125
formula, 107, 114, 142, 143, 145
freezing, vii, 2, 7, 13, 22, 35, 49
friction, 36, 40
frost, 123
fructose, 119, 123, 124, 125
functional food, vii, 1, 120
fungi, 105

G

gastrointestinal tract, viii, 103, 104, 105, 106, 116, 119, 124, 130, 132, 133
gel, vii, 2, 11, 14, 15, 19, 21, 30, 40, 54
gel formation, 54
genes, 109, 111
genetic information, 110

genus, 105, 116, 117, 128, 140
Germany, 7, 144
gestational age, 107
glucagon, 142, 144
glucose, 113, 114, 119, 121, 122, 123, 124, 130, 138, 141, 142
gluten, 114
GPA, 54
graph, 17, 18
GRAS, 124
growth, 3, 105, 106, 111, 112, 118, 121, 125, 127, 130, 131, 137
guidelines, 120

H

habitat, 104
harbors, 117
harmful effects, 110
health, vii, 3, 4, 47, 48, 107, 109, 118, 119, 120, 121, 126, 127, 130, 133, 134, 135, 136
health care, 136
health effects, 48, 118, 130
health promotion, 47
hemoglobin, 122, 131
hepatocellular carcinoma, 131
hepatocytes, 114
high density lipoprotein, 115
high fat, 109
homeostasis, 113, 115, 118, 122
homogeneity, 38, 45
hormone, 129
hormones, 128
host, 104, 107, 108, 109, 110, 112, 116, 118, 119, 127, 128, 130, 131, 134, 135
hue, 8
human, 4, 105, 107, 108, 112, 120, 121, 123, 124, 128, 130, 134, 135, 136, 137, 138, 140, 143, 144
human body, 107
human health, 120, 121, 123, 128, 130, 135, 140
human subjects, 137, 143
Hunter, 108
hunter-gatherers, 108, 136
hydrocolloids, vii, 2, 4, 5, 9, 10, 11, 12, 13, 14, 17, 18, 19, 20, 22, 26, 27, 29, 30, 32, 33, 36, 46
hydrogen, 106, 129
hydrolysate, viii
hydrolysis, viii, 47, 111, 119, 124, 125
hygiene, 107
hypercholesterolemia, 133
hyperlipidemia, 127
hypersensitivity, 116

I

IBD, 113, 115, 116, 137
ideal, 47
identification, 6, 10, 110, 130
ileum, 104
immune function, 126, 127, 130, 138
immune response, 110, 113, 116, 127
immune system, viii, 103, 106, 107, 114, 138
immunity, 110
immunogenicity, 107
immunoglobulins, 107
immunomodulation, 127
immunomodulatory, 127
improvements, 109, 113, 114, 129
in vitro, 125
in vivo, ix, 54, 104, 134
incidence, 114
individuals, 44, 108, 110, 113, 114, 121, 127, 128, 130, 131, 132, 133
industries, 123
industry, ix, 104, 123, 124, 133, 134
infancy, 139
infants, 139
infection, 116, 127
infectious agents, 106
inflammation, 110, 116, 117, 127, 131, 132, 140, 141
inflammatory bowel disease, 113, 115, 138, 139
ingestion, 3

ingredients, 4, 5, 6, 9, 14, 18, 24, 29, 30, 33, 37, 46, 107, 119, 121, 141
INL, vii, 1, 3, 4, 5, 6, 10, 11, 12, 13, 14, 15, 16, 17, 18, 19, 20, 21, 22, 23, 24, 25, 26, 27, 28, 29, 30, 31, 32, 33, 34, 35, 36, 37, 38, 39, 41, 43, 46, 47
inositol, 123
Instron, 7
insulin, 108, 110, 113, 114, 115, 122, 127, 128, 130, 131, 132, 136, 137, 138, 141, 142
insulin resistance, 108, 110, 113, 114, 115, 127, 131, 132, 136, 138, 142
insulin sensitivity, 128, 130, 132
insulin signaling, 114
integrity, 127, 131
interdependence, 10
interface, 36
internal consistency, viii, 2, 44
intervention, 112, 120, 127, 129, 131, 133, 135, 137, 138, 142, 143
intestinal epithelium, viii, 103, 106, 133
intestinal flora, 105, 133
intestinal tract, 105, 111
intestine, 110, 127, 134
inulin, vii, viii, ix, 16, 17, 20, 22, 25, 26, 27, 28, 32, 33, 34, 35, 37, 39, 43, 47, 49, 50, 51, 52, 53, 54, 103, 104, 107, 119, 120, 121, 122, 123, 124, 125, 126, 127, 128, 129, 130, 131, 132, 133, 134, 138, 140, 141, 142, 143, 144, 145
Inulins, vii, 53
ions, 17
iron, 105, 133, 143
irritable bowel syndrome, 116, 139
Israel, 124
Italy, 31

J

Japan, 124
jejunum, 104
Jerusalem artichoke, 119, 124, 143, 145

K

kidney, 117, 139, 140
kinetics, 129

L

lactobacillus, 53
Lactobacillus, 104, 105, 107, 112, 114, 116, 137, 140, 141, 142
large intestine, 104, 111
LDL, 113, 122, 132, 133
lead, 15, 18, 36, 125, 131
lean body mass, 130
life cycle, 125
light, 7, 21, 23
light beam, 21
light scattering, 21, 23
linear polymers, 124
lipid metabolism, 126, 133, 142
lipid peroxidation, 128
lipids, 112, 133
lipoproteins, 132
liquids, 48
liver, viii, 103, 108, 115, 118, 131, 132, 138
liver cells, 132
liver disease, viii, 103, 108, 115, 118, 132, 138
liver failure, 131
low temperatures, 123, 124
low-grade inflammation, 128
LSD, 10, 16
lumen, 117
lung disease, 117
Luo, 142
lymphoid tissue, 110

M

macromolecules, 3
macronutrients, 112
macrophages, 111, 131
magnesium, 122, 132, 144

magnitude, 14, 27, 36
majority, 123
management, 54, 121, 122
mannitol, 122, 123
mass, 106, 110, 122, 130
materials, 15
matrix, 12, 14, 15, 17, 46, 51
matter, 128
measurement, 11
measurements, 10, 22, 24, 28, 44, 54
meat, ix, 104, 111, 136
mechanical properties, 8, 49, 50
media, 112
medical, 120
Mediterranean, 3
mellitus, 114
melting, 39, 41, 51, 52
meta-analysis, 113, 115, 128, 129, 130, 137, 138, 142
Metabolic, 122, 131, 136, 140
metabolic disorders, 110, 127, 128, 134
metabolic pathways, 118
metabolic syndrome, 131, 138
metabolism, viii, 31, 103, 104, 105, 111, 121, 122, 127, 130, 137, 138, 142
metabolites, 116, 117, 127, 133, 137
meter, 8
methodology, 47
microbial cells, 104
microbial communities, 113
microbial community, 104, 105
microbiota, viii, 103, 104, 105, 106, 107, 108, 109, 110, 111, 112, 113, 114, 115, 116, 117, 118, 120, 121, 122, 127, 130, 134, 135, 136, 137, 138, 139, 140, 142, 143
microorganisms, 3, 104, 105, 106, 107, 109, 115, 116, 127, 140
microstructure, 15, 41, 49
mineralization, 132, 144, 145
MLT, 145
models, 109, 112, 132, 133
modifications, 121
modulus, vii, 2, 8, 11, 13, 14, 17, 20
moisture, 7, 51
molecular weight, 124
molecules, 13, 48, 124, 125, 127, 129
monosaccharide, 119, 123, 125
monozygotic twins, 114
mortality, 140
mucin, 115, 127
mucosa, 105, 107, 110
mucus, 106
multiplication, 126

N

Na^+, 16
NaCl, 5
native starches, 50
natural food, 3, 120
neurophysiology, 4
neutral, 30, 125
New Zealand, 124
nitrogen, 7, 111, 117
North America, 119
Norway, 124
nucleation, 19
nucleic acid, 125
nutrient, 111, 136
nutrients, 104, 105, 127

O

obesity, viii, 103, 108, 109, 113, 118, 126, 128, 131, 134, 136, 137, 138, 142
oil, viii, 2, 3, 15, 21, 23, 24, 30, 35, 36, 37, 46, 48, 50, 51, 52, 123
oleic acid, 3, 31
oligomers, 124
oligosaccharide, 123, 140, 142
olive oil, vii, 1, 2, 3, 16, 17, 20, 21, 22, 25, 26, 27, 28, 30, 31, 32, 33, 34, 35, 37, 38, 39, 43, 46, 48, 49, 51, 52, 53
optical microscopy, 15, 40
optical properties, 21, 52
optimization, 47
organ, 104, 110

organism, 106
oscillation, 11, 14
osmotic pressure, 123
osteoporosis, 132
overlap, 37
overproduction, 132
overweight, 108, 113, 128, 129, 133, 137, 138, 142
oxidative stress, 116, 127
oxygen, 128

P

pain, 116
palate, 38
pancreas, 114
participants, 129, 132
pasta, 142, 143
pathogenesis, 130, 132
pathogens, 110, 112, 113
pathophysiology, 114
pathways, 115, 116
PCA, 5, 10
PCR, 125, 139
peptide, 106, 122, 128, 130, 142, 144
peptides, 121, 122, 143
permeability, 110, 117, 121, 122, 131, 142
pH, viii, 49, 105, 106, 111, 117, 122, 125, 127, 133, 142
pharmaceutical, 123, 144
pharmacology, 134
pharmacotherapy, 117
phenolic compounds, 31, 38, 106
phenotype, 108
phenotypes, 113, 139, 140
phosphate, 117
phylum, 105, 109
physiology, 104, 106, 135
pilot study, 142, 144
placebo, 129, 142, 143
plants, vii, 111, 123, 125
plasma levels, 131
playing, 106
polydispersity, 124
polymer, 123
polymerase chain reaction, 125
polymerization, 119, 123, 124, 125
polymers, 123
polyphenols, 3, 112, 137
polysaccharide, 15, 17
polysaccharides, vii, 3, 31, 51, 111, 119, 123, 125
polysacharides, viii
polyunsaturated fatty acids, 3
population, 108, 109, 111, 112, 116, 131, 137, 142
portal hypertension, 131
positive correlation, 114
potassium, 117
potato, 9, 12, 21, 22, 24, 30, 31, 37, 50, 51
prebiotic, vii, ix, 3, 47, 104, 107, 118, 119, 120, 121, 123, 128, 129, 131, 132, 133, 134, 138, 140, 141, 142, 143, 144, 145
precipitation, 125
preparation, 7, 41
prevention, 47, 132
principal component analysis, viii, 2, 10
Principal Components Analysis, 40
probiotic, 52, 107, 132, 133, 140
probiotics, 107, 118, 120, 129, 131, 135, 138, 140, 141
producers, 115, 120
professionals, 120
pro-inflammatory, 108, 110, 111, 114, 121, 131
project, 54
proliferation, viii, 103, 106, 112, 126, 130
protection, viii, 103, 105, 106, 123, 131, 134
protective role, 115
proteins, 3, 4, 110
psychology, 4
public health, 121
purity, 5

R

radar, 24
radiotherapy, 142
rancid, 31
raw materials, 125
reaction temperature, 126
reactions, 104
reactive oxygen, 128, 131
receptors, 31, 38, 105, 110, 114, 115
recognition, 107, 115
recommendations, 121, 126
reducing sugars, 31
redundancy, 43
reinforcement, 18
relevance, 135
reliability, 5, 10, 44
remission, 113, 116
requirements, 118
researchers, 123, 132
residues, 119
resistance, 11, 13, 127, 131
response, 14, 46, 47, 110, 121, 138
restrictions, 117
reticulum, 127
rheology, 12, 46, 50, 52
ribosomal RNA, 109
risk, 3, 131
rodents, 109
root, 51
roughness, 29, 36, 38, 41
routes, 117
rural areas, 111

S

sacharides, viii
safety, 48
saliva, 54
Samsung, 7
saturation, 8
scanning electron microscopy, 35
scattering, 21
secretion, 128, 130, 132
self-assessment, 129
self-reports, 129
sensations, 29, 36, 38, 40, 54
serine, 114
serum, 122, 132, 133, 141, 142
sex, 109
shear, vii, 1, 7, 11, 18, 40, 48, 50
shear data, vii, 2
showing, 16, 44, 107, 114
side effects, 121
signal transduction, 136
signals, 110, 114, 129
signs, 107
skeleton, 122, 144
skin, 104
small intestine, 127
smoothness, 29, 30, 41
sodium, 4, 49, 51, 106
software, 10
solubility, 15, 125
solution, viii, 125
solvents, 3
Spain, 1, 5, 7, 31
species, 104, 105, 108, 109, 114, 115, 116, 117, 128, 130, 131, 140, 141
spectroscopy, 11
speculation, 15
stability, 14, 18, 19, 30, 35, 36, 46, 49, 50, 51, 118
stabilization, 123
starch, 4, 12, 13, 15, 17, 18, 19, 30, 31, 38, 49, 50, 51, 53, 106, 111, 112
starch polysaccharides, 106
state, 109, 110, 113, 114, 115
states, viii, 103, 108, 113, 115, 118
steel, 7
sterilisation, viii
stimulation, 115, 143
stimulus, 4, 108
stomach, 104, 128
storage, vii, 2, 7, 8, 11, 13, 14, 17, 49, 50, 125
stress, 7, 11, 13, 15, 48, 109, 123, 132
structural changes, 11, 35